Pesticides in Household, Structural and Residential Pest Management

ACS SYMPOSIUM SERIES **1015**

Pesticides in Household, Structural and Residential Pest Management

Chris J. Peterson, Editor
USDA Forest Service

Daniel M. Stout II, Editor
U.S. Environmental Protection Agency

Sponsored by the
ACS Division of Agrochemicals

American Chemical Society, Washington DC

Library of Congress Cataloging-in-Publication Data

Pesticides in household, structural, and residential pest management / Chris J. Peterson, Daniel M. Stout II, editors ; sponsored by the ACS Division of Agrochemicals.

p. cm. -- (ACS symposium series ; 1015)

Includes bibliographical references and index.

ISBN 978-0-8412-6967-5 (alk. paper)

1. Insecticides. 2. Insects as carriers of disease. 3. Pesticides--Formulation. 4. Household pests--Control. I. Peterson, Chris J. II. Stout, Daniel M.

TP248.P47P496 2009

614.4'32--dc22

2009031668

The paper used in this publication meets the minimum requirements of American National Standard for Information Sciences—Permanence of Paper for Printed Library Materials, ANSI Z39.48–1984.

Distributed by Oxford University Press

PRINTED IN THE UNITED STATES OF AMERICA

Foreword

The ACS Symposium Series was first published in 1974 to provide a mechanism for publishing symposia quickly in book form. The purpose of the series is to publish timely, comprehensive books developed from the ACS sponsored symposia based on current scientific research. Occasionally, books are developed from symposia sponsored by other organizations when the topic is of keen interest to the chemistry audience.

Before agreeing to publish a book, the proposed table of contents is reviewed for appropriate and comprehensive coverage and for interest to the audience. Some papers may be excluded to better focus the book; others may be added to provide comprehensiveness. When appropriate, overview or introductory chapters are added. Drafts of chapters are peer-reviewed prior to final acceptance or rejection, and manuscripts are prepared in camera-ready format.

As a rule, only original research papers and original review papers are included in the volumes. Verbatim reproductions of previous published papers are not accepted.

ACS Books Department

Contents

Indexes

Chapter 1

The Chemistry of Household, Structural and Residential Insect Management

Chris J. Peterson[1] and Dan Stout, II[2]

[1]USDA-Forest Service, Insects, Diseases and Invasive Plants Unit, 201 Lincoln Green, Starkville, MS 39759
[2]US Environmental Protection Agency, Human Exposure and Atmospheric Science Division, 109 T.W Alexander Dr., Durham, NC 27711

Most, but not all, of the active ingredients used to control household and structural insect pests were developed for agricultural uses before entering the household and structural market. Therefore we are using compounds in situations for which they were not originally developed. Exceptions to this are the termite, cockroach and ant baits, such as those containing hexaflumuron, noviflumuron, sulfluramid or hydramethylnon. Most other products, however, are formulations of agricultural products or derivatives of products originally developed for agriculture.

The usual Rogue's Gallery of household and structural insect pests are rarely, if ever, pests in agricultural settings. Termites are not pests of living vegetation in the United States, nor are mosquitoes, fleas, cockroaches or bedbugs. Other than the occasional and incidental invader or garden pest, agricultural pests are largely unknown to the homeowner. Pantry pests such as the Indian meal moth and the red flour beetle, however, are pests in agricultural grain storage and pests of pets can also occur in livestock.

Household, structural and residential insect pest management touches all of our lives. Pesticides, specifically insecticides, are used in virtually every environment we encounter in our daily activities. In our workplaces, from the farm to the business office, in schools and daycares, as well as in our homes, insecticides are applied to control peridomestic insect pests. Information regarding the use of insecticides in our living environment is incomplete, but surveys suggest that about 75% of American households used pesticides in the past year (*1*). Depending on age, gender, work and family composition, people typically spend about 90% of their day in indoor locations (*2*). Pesticide use in and around our daily environments potentially places humans in intimate association with chemicals used to control structural pests.

Human competition with insects for food, shelter and health has been an historic conflict. However, a schizm between our use of pesticides and our perception of ther risks has slowly evolved. Pesticides regulated by the state and federal government are approved for use and deemed safe and efficacious; however, society often negatively perceives their use to control insects, diseases and other pests as having high inherent risk to our health and the environment. In contrast there is an indirect acknowledgment of the need to control pests expressed by the ready availability of consumer-use products and the services of pest control professionals.

The issues of perception, risk and necessity of control procedures might be best exemplified in the public housing arena. The negative effect of cohabitating with cockroaches is generally acknowledged due to the recognition of cockroach feces and chitin as an asthma trigger. High cockroach populations in public housing and the related health issues has resulted in a history of elevated pesticide usage. Of late, substantial state and federal resources have been expended to introduce intregrated pest managmemnt principals to reduce or eliminate pesticide use in public housing. Grassroots or community-based efforts, inspired by the desire to transition away from the more conventional spray approach to reduce pesticide exposure, have further impacted pesticide use in public housing. In contrast, prophylactic treatments in and around private dwellings to control a wide variety of insect pests continue, even when there is no evidence of pest infestation. In fact, an estimated 78 million U.S. households (*3*) spent nearly 1.3 billion dollars to purchase insecticides and applied 888 million pounds of active ingredient (*1*).

The development of new household and structural insect management products continues apace. It is unlikely that any one product or process will be the "silver bullet" of household and structural pest control. What will in reality happen is that pest management professionals and homeowners will have an ever-larger kit of management tools available to them, and each tool is highly effective in its designed capacity. Seeing, as we did above, that the home and garden insect control market is as potentially lucrative as the agricultural market, it is in the interests of manufacturers to continue to develop new products that can meet changing market demands and the regulatory environment. With this in mind, the safety and efficacy of these products in household, structural and residential situations will continually need to be evaluated and merged with ongoing societal concerns at both the regulatory and community levels.

The ACS Symposium Series has previously produced an excellent title, although now fifteen years old, related to this subject (*4*). This book continues the interest by focusing on the development of new household and structural insect management products. This volume examines several phases of the process of discovering, developing, using and monitoring for the insect management tools used in and around the home. Several chapters address pesticide efficacy in controlling different species of termites. Here chapters discuss the discovery of new active ingredients (including natural products), the evaluation of efficacy, and issues relating to liquid formulations and baits are discussed, as are biological and environmental factors that affect efficacy and longevity. A chapter focuses on insecticide mode of action as it relates to a novel compound and its implications for controlling household pests. Reflective of ongoing interests in residential integrated pest management, a chapter is dedicated to exploring the least-toxic approaches for controlling insect pests. Two chapters examine potential human exposure associated with insecticidal control of ectoparasites on companion animals, and their transfer to hands during contact, and the potential role of dogs in transporting pesticide residues into homes.

The intent of this book is to present a broad spectrum of topics associated with residential control and continue to build on the topic through the ACS Symposium Series.

References

1. Kiely,T.; Donaldson, D.; Grube, A. Pesticide Industry Sales and Usage, 2000 and 2001 market Estimates. United States Environmental Protection Agency, Washington, DC. EPA-733-R-04-001, 2004; 48 pp.
2. Klepsis N. E.; Nelson W. C., Ott W. R.; Robinson J. P.; Tsang A. M.; Switzer P.; Behar J. V.; Hern S. C.; Engelmann W. H. The National Human Activity Pattern Survey (NHAPS): a resource for assessing exposure to environmental pollutants. *J. Expo. Anal. Environ. Epidmiol.* **2001,** *11,* 231–252.
3. U.S. EPA. Pesticides industry sales and usage report. 1998 and 1999 market estimates. Office of Prevention, Pesticides and Toxic Substances, Washington, DC: USA, 2002.
4. Racke, K. D.; Leslie, A. R., Eds. *Pesticides in Urban Environments*; ACS Symposium Series 522. American Chemical Society, Washington DC, 1993; pp. 282–295.

Chapter 2

Amyris and Siam-wood Essential Oils: Insect Activity of Sesquiterpenes

Gretchen E. Paluch, Junwei Zhu[1], Lyric Bartholomay, and Joel R. Coats

Department of Entomology, Iowa State University, Ames, IA, 50011
[1]USDA-ARS, Agroecosystem Management Research, 305 Plant Industry Building, East Campus, University of Nebraska, Lincoln, NE 68583-0938

Recent investigations on the sesquiterpene-rich Amyris (*Amyris balsamifera* L.) and Siam-wood (*Fokienia hodginsii* L.) essential oils revealed significant arthropod repellency and toxicity responses. Amyris essential oil and one of its major components, elemol, were evaluated in laboratory bioassays and identified as effective mosquito repellents, specifically characterized by high levels of contact and minimal spatial repellency. Mosquito responses to catnip (*Nepeta cataria* L.) essential oil are characterized with high spatial activity, but lack significant contact repellency. Sampling within the static-air bioassay chamber with solid-phase microextraction provided measurements of the relative concentration and distribution of volatiles. These results supported the differences observed in repellency between essential oil treatments. Essential oil mixtures containing both spatial (catnip) and contact (Amyris) repellents were made and showed high levels of residual control via both modes of action. Siam-wood essential oil scored high in both spatial and contact efficacy against mosquitoes. Observations during this study included signs of toxicity. Two of the primary components of Siam-wood essential oil were tested for 24-hour house fly (*Musca domestica* L.) topical mortality. Trans-nerolidol and fokienol were found to possess similar insecticidal activity (topical LD_{50} values ranged from 0.17-0.21 μmol/fly). Amyris essential oil was selected for additional testing with brown dog ticks (*Rhipicephalus sanguineus* Latreille) in a 'barrier' repellency assay. Individuals were observed repeatedly avoiding and moving away from surfaces treated with Amyris essential oil.

Introduction

Nature holds a diversity of terpenoid structures, and the functionality of these compounds is still poorly understood. Only a small number actually serve a primary metabolic function (ex. carotenoids, sterols, etc.). In the 1970s, researchers started to identify other terpene bioactivities including toxicity, attraction, and repellency (*1*). The challenges today still include the characterization of terpene function, but also improvement of our understanding of their ecological roles. A variety of living organisms are known to utilize terpenes for coordinating antagonistic and beneficial interactions, such as inter- and intraspecific communication, and defense (*2*).

Terpenoid compounds are classified into groupings based on the number of isoprene units: hemiterpenes C5, monoterpenes C10, sesquiterpenes C15, diterpenes C20, sesterterpenes C25, triterpenes C30, tetraterpenes C40, and polyterpenes (terpene polymers). In plants, terpene biosynthesis pathways are either via the formation of a mevalonic acid intermediate or the pyruvate pathway. Mono-, sesqui-, and diterpenes are formed by continual addition of 5-carbon units, whereas other larger terpenes require joining of large carbon units, e.g. two sesquiterpenes to form a triterpene.

Bioactivity of Sesquiterpenes

Sesquiterpenes are produced in a number of plant families and appear in different concentrations in the essential oil composition. In many of these cases sesquiterpenoids make up only a small percentage of the essential oil blend, however there are examples of oils containing large amounts of these compounds with similar ring structures and specific functional groups. There is evidence of essential oils, and the actual plant tissues (heartwood, bark, leaves, etc.), containing sesquiterpenes with alcohol, aldehyde, and acid moieties, possessing high levels of insecticidal or repellent activity. The essential oil obtained from the bark of *Goniothalamus uvariodes* King, a small tree endemic to Borneo, is one example. Both the bark and leaves from this plant are used by several local groups including the Kedayan and Iban communities in Sarawak and the Sungai in Sabah as an insect repellent. The chemical constituents of the bark includes sufficient amounts of nerolidol (5.2%), α-eudesmol (5.6%), hedycaryol (13.6%), γ-eudesmol (16.0%), and β-eudesmol (31.5%) (*3*). These compounds and other closely related structures (farnesane, eudesmane, eremophilane, and elemene derivatives) appear in other reports detailing insect response to essential oils.

Several eudesmol isomers, and a eudesmane sesquiterpene acid and methyl ester derivatives were isolated from *Callitris glaucophylla* Thompson et Johnson and identified as termite repellents (*4*). The *Cryptomeria japonica* (L. f.) D. Don essential oil contains elemol as its major component (18.2%), and was recently identified as a repellent to silverfish (*5*). Another interesting study investigated the essential oil composition of *C. japonica* cultivars that varied in susceptibility to the *Cryptomeria* bark borer (*Semanstus japonicus* Lacordaire). Attractant and repellent responses of the *Cryptomeria* bark borer were used to assay select chemical components of the essential oils, and quantitative comparisons were made across the different cultivars. There were notable differences in the essential oil compositions of the resistant and susceptible cultivars, with the bark oils showing great diversity in structures and amounts of terpene hydrocarbons in particular, pinene (16-52%), limonene (7-12%), and δ-cadinene (4-8%). Many of the terpene hydrocarbons, e.g. β-pinene, camphene, sabinene, β-phellandrene, β-caryophyllene, and longifolene, were found to be attractants for the *Cryptomeria* bark borer. Four compounds were found to occur in significantly higher levels in the resistant cultivars and identified as repellents in the laboratory bioassay. These included three oxygenated sesquiterpenes α-terpineol, nerolidol, and β-eudesmol (*6*).

Callicarpenal and intermedeol were isolated from the American beautyberry bush (*Callicarpa americana* L.) and recently tested for insect activity. Researchers used a finger tip climbing assay and found both to be effective tick repellents. At an application rate of 155 nmole/cm^2 deer tick (*Ixodes scapularis* Say) nymphs were repelled 98 and 96%, respectively. These compounds were compared with commercial standard N,N-diethyl-m-toluamide (DEET) , and there was no significant difference with DEET (callicarpenal, EC_{50} 14.2 nmol/cm^2; intermedeol, EC_{50} 17.4 nmol/cm^2; DEET, EC_{50} 23.9 nmol/cm^2) (*7*).

Another collection of sesquiterpenoids from the heartwood of the Alaska yellow cedar (*Chamaecypars nootkatensis* D. Don), include nootkatone and valencene-13-ol. Both of these compounds were just as repellent to *I. scapularis* as DEET (nootkatone, RC_{50} 0.0458% wt/vol solution; valencene-13-ol, RC_{50} 0.0712% wt/vol solution; DEET, RC_{50} 0.0728% wt/vol solution) (*8*).

Amyris Essential Oil

West Indian sandalwood or Amyris oil (*Amyris balsamifera* L.) is produced from the heartwood of a small tree (3-6 m, 75-150 DBH) in the Rutaceae. Some of the identifying features of this tree include three to seven ovate, opposite and compound leaflets, white flowers in lateral clusters, and a black drupe fruit. Trees are described as having a smooth grayish bark, with a rounded crown of aromatic foliage. Its distribution is mostly limited to the Caribbean islands, but is also found in some South American countries. Amyris is also referred to as *bois chandelle* (candlewood) in Haiti, torchwood in Jamaica, *tigua* in Venezuela, but in the United States as Amyris, balsam amyris, or West Indian sandalwood. Interestingly, this species is not closely related to the other sandalwood (e.g. Indian or Australian sandalwoods), which are highly valued, wood-scented essential oils derived from trees in the Santalales. The sandalwood oils and other byproducts (including incense, pastes, and wood-carvings) have a rich history of being used in religious and social ceremonies.

Some other common uses for the Amyris heartwoods have included torches, firewood, fence posts, and ancient wood-carvings mosaics (*9*). This is not surprising considering the soft-quality of the heartwood and its use in carving. Also, there are studies citing the antimicrobial activity of Amyris extracts. Amyris essential oil is an effective inhibitor of *Klebsiella pneumonia* growth, and minimally effective against *Staphylococcus aureus* (gram-positive), *Escherichia coli* (gram-negative), and *Pseudomonas aeruginosa* (*10*). Such properties would no doubt be beneficial for maintaining the integrity of the wood in several of the uses listed above.

In most regions where Amyris is commercially grown, it is used for essential oil production. Steam distillation is estimated to yield 2-4%, depending on the portions of wood used. The essential oil is a viscous amber liquid composed mostly of oxygenated sesquiterpenes (80%) and sesquiterpene hydrocarbons (20%). Its woodsy scent is used in perfumery, soaps, and cosmetics and is also believed to be used by the cosmetic and perfume industries to dilute more expensive sandalwood oils such as that from East Indian sandalwood, *Santalum album* L. (*11*). There are also pharmaceutical and nutriceutical benefits from Amyris chemistries. Anti-mutagenic activity has been shown with β-eudesmol, one of the primary components. This compound suppressed SOS-inducing activity of furylfuramide, in addition to suppression of gene expression (ID_{50} 0.09 μmol/ml) in *Salmonella typhimurium* TA1535/pSK1002 with the furylfuramide mutagen 2-(2-furyl)-3-(5-nitro-2-furyl)acrylamide. Additional suppression activity was seen against the Trp-P-1 mutagen 3-amino-1,4-dimethyl-5H-pyridol[4,3-b]indole (*12*).

Previous studies in the Pesticide Toxicology Laboratory at Iowa State University, Ames, IA identified the repellent activity of Amyris essential oil against mosquitoes (*13*). Amyris was one of forty essential oils recently screened for repellency of *Aedes*, *Anopheles*, and *Culex* spp. mosquitoes using the human-bait technique (*14, 15*). The Amyris essential oil formulation provided a 480-minute protection period against *Anopheles* and *Culex* and 240 minutes for *Aedes*. Percentages of landing and biting mosquitoes reported was also low (*Anopheles*, 0% landing and biting; *Culex,* 0% landing and biting; *Aedes,* 9.6% landing and 0.8% biting). These levels were comparable to the Bayrepel and DEET formulations (*16*). Studies with Amyris essential oil as a potential mosquito larvicide were conducted using the yellow fever mosquito, (*Aedes aegypti* L.). With fresh preparations, researchers found 100% mortality of the mosquito larvae at 6 h following application, at a rate of 50 ppm (*17*). Efficacy following storage of this preparation showed that it was not effective after 1 week in a dark environment.

Siam–wood Essential Oil

Siam-wood (*Fokienia hodginsii* L.), which is also known as Vietnamese pemou, produces a highly prized oil from the heartwood in the Cupressaceae. These cypress trees are the only living species in the genus *Fokienia* and are adapted to growing at higher altitudes (600-1800m) in regions of Southern China, Northern Lao PDR, and Vietnam (*18*). Some of the people in these regions, such as the Greater Annamites, utilize the wood for housing and furniture construction. This is due to the longevity of the wood and its ability to handle many climatic factors and resist insect injury. The essential oil is extracted from the stumps and roots. Constituents of the essential oil were reexamined by Weyerstahl et al., and they found only sesquiterpenes. The major components identified were (E)-nerolidol (34.8%) and fokienol (25.7%); minor components were multiple cadinene isomers (6.5%), eudesmol isomers (7.4%), α-cadinol (1.9%) and dauca-8(14),11-dien-9-ol (3.1%) (*19*). There is limited literature available on the insect activity of Siam-wood extracts. Only one citation was found that mentioned that the wood is resistant to termites and moths (*19*).

The intent of this study was to characterize the bioactivity of two sesquiterpene-rich essential oils, Amyris and Siam-wood. In the initial screening trials, both oils showed evidence of repellency against a mosquito (*Ae. aegypti*). One area of particular interest was observation of residual repellency effects (including both contact and spatial repellency), which were supported by the relative concentration of volatiles measured inside the bioassay chambers. These essential oils were evaluated against actives contained in commercial natural products, and then incorporated into mixtures to test for improvements of natural product residual efficacy. The results of this study show that Amyris and Siam-wood significantly repel arthropods, are superior to other natural products in today's market, and could potentially be utilized to improve residual control in repellent formulations.

Materials and Methods

Mosquito Repellency Bioassay

Bioassays were conducted in a static-air apparatus (9 x 60-cm section of glass tubing) at a controlled temperature of 26°C. Yellow fever mosquitoes (*Aedes aegypti*), a Costa Rican strain, were from an established laboratory colony in the Iowa State University, Medical Entomology Laboratory, Ames, Iowa. Eggs were hatched in deoxygenated water, and larvae were fed Tetramin fish food (Melle, Germany). Pupae were sorted from the larvae and placed in paper cups with mesh lids until emergence. Newly emerged adults were fed a 10% (0.3 M) sucrose solution and aged for at least 5-days before testing. Incubator conditions were set at 60% relative humidity and held at 27°C. Only female mosquitoes were used in the testing.

Essential oils and mixtures included catnip (*Nepeta cataria* L.) oil, which was produced from a steam distillation in the laboratory (*20*). Amyris oil was purchased from Sigma Aldrich, St. Louis, Missouri; Siam-wood essential oil was purchased from Oshadhi, Petaluma, California. Elemol, a sesquiterpene found in both Amyris and Siam-wood essential oil, was purified from a crude commercial source (Augustus Oils, New Hampshire, England) using column chromatography techniques with silica gel. Several of the commercial repellent active compounds were available for purchase: DEET, citronella oil, 2-

undecanone, and cis/trans p-menthane-3,8-diol (Sigma Aldrich, St. Louis, Missouri).

Test solutions were made up in a carrier solvent (either acetone or hexane), applied to 9-cm diameter round filter papers (63.6 cm^2), and then the solvent was evaporated off prior to testing. The resulting rate of exposure was 78.6 $\mu g/cm^2$. Treated filter papers were placed inside the lids of 9-cm glass petri dishes, and the dishes were placed over the ends of the glass chamber. A group of 20 female mosquitoes were anaesthetized with CO_2 and introduced through a 2-cm hole drilled at the midpoint of the chamber. Mosquito distribution inside the static-air choice-test apparatus was observed over a total of 360-minutes. The experimental design was a completely randomized design using three replications of each treatment. Data generated by this study was used to examine two measures of mosquito repellency, **percentage (spatial) repellency** and **contact repellency**. Percentage repellency was calculated with the following formula to provide an indication of spatial repellency:

Percentage Repellency = ((Number of Individuals in Untreated Half - Number of Individuals in Treated Half) / 20) × 100

Contact repellency was defined in this assay as 100% avoidance of the treated filter paper (no contact) throughout the 360 minute observation period. The resulting contact repellency was compared with control treatments, using Fisher's Exact Test.

Collection of Volatiles Using Solid-Phase Microextraction

Relative concentrations of volatiles were sampled inside the static-air glass apparatus used in the repellency bioassays. Test solutions were applied to filter papers at a rate of 78.6 $\mu g/cm^2$ and then enclosed in the system. Catnip essential oil, elemol, and DEET were selected, based on the differences in mosquito repellency (contact vs. spatial activity) observed in the previous bioassay. Temperature and light were held constant throughout the study. Solid-phase microextraction (SPME) field samplers containing a PDMS fiber (Supelco, St. Louis, Missouri) were conditioned in a GC inlet held at 250°C for 30 minutes before sampling. Holes were drilled in the center of equally-spaced quadrants of the static-air chamber and covered with a small amount of parafilm, to allow placement of the four SPME fibers in each volatile sampling replicate. Prior to the start of the study, static-air chambers were sampled with SPME fibers and identified a minimal level of background contamination.

SPME fibers were exposed inside the treated chambers for one of two 15-minute time periods; collection of volatiles was conducted immediately following treatment (0-15 min.), or 15 minutes after treatment (15-30 min.). Volatile samples were replicated three times for each test solution and time period. Relative concentrations of volatile samples were measured by GC-FID. Quantitative standards were made up for DEET (Sigma Aldrich), as well as elemol (≥ 80%), *Z,E*-nepetalactone (≥ 90%), and *E,Z*-nepetalactone (≥ 90%), which were purified in the laboratory by column chromatography. Theoretical vapor pressures were calculated using ACD/Lab Boiling Point software, Version 8.0.

House Fly Toxicity Test

Toxicity bioassays were performed with adult house flies (*Musca domestica* L.), from an established laboratory colony in the Iowa State University, Pesticide Toxicology Laboratory, Ames, IA. Individuals were chilled on a cooled surface and dosed with one μl of test solution on the ventral abdominal surface. Test solutions consisted of five different concentrations of the active ingredient in an acetone solvent along with an acetone-only control, dispensed using a topical applicator (Model PB-600, Hamilton Co., Inc., Whittier, California). Each concentration was applied to a population of 10 house flies and then placed in a screen-covered glass mason jar containing a cotton wick soaked in a saturated sucrose solution. Mortality was recorded after 24-hours. All treatments were replicated three times.

Tick Repellency Bioassay

Tick responses to candidate repellent essential oils and compounds were evaluated in a climbing arena. Positive controls consisted of DEET and a 20% pyrethrum solution (Sigma Aldrich). Brown dog ticks (*Rhipicephalus sanguineus* Latreille) were purchased from EL Lab, Soquel, California. Four individuals were placed in a glass Petri dish arena (area of 10.2 cm^2) surrounded by water, maintained at 23-24°C. In the center of the arena, a braided cotton wick was suspended. Treatments were made up as solutions in acetone and applied evenly across a "barrier", designed at 2.54 cm from the bottom of the arena. The solvent was allowed to evaporate off the cotton wick (1-2 minutes) prior to the start of the test period. Ticks were allotted 60 minutes to search the arena and begin climbing behavior. The total number of ticks that attempted to climb the cotton wick was recorded. Individuals that passed the treated barrier were removed from the arena and recorded. If a tick approached the chemical barrier and either circled or turned around, the activity was noted and then the individual was allowed to continue movement in the arena until the 60 minutes had concluded. Five replications were completed for each treatment.

Results

Results for Amyris essential oil and for a mixture (1:1), containing a potent spatial repellent, catnip essential oil, are shown in Table 1. The difference between Amyris and catnip oils can be seen in the comparison of their percentage repellency values (measure of spatial repellency) and avoidance frequency (contact repellency). Amyris yielded a significant degree of spatial repellency compared to the control, but this percentage repellency value was lower than for the catnip oil. There was also a noticeable difference in avoidance frequency of Amyris and catnip. Amyris avoidance frequency accumulated over the 3-hour test period was 0.97, i.e. only 1 mosquito came in contact with the treated filter paper. The Amyris and catnip essential oil mixture resulted in significant levels of both spatial repellency and contact repellency.

Table 1. The 15-minute spatial repellency and 3-hour contact repellency of yellow fever mosquitoes (*Aedes aegypti*) exposed to 78.6 μg/cm² rate of Amyris and catnip essential oils and mixtures (1:1) in the static-air repellency chamber.

Treatment	*Percentage Repellency* [a]	*Std. Dev.*	*Avoidance Frequency* [c]	*Contact Rep.* [d] *(P value)*
Catnip Essential Oil	77.7*	14	0.19	0.218
Amyris Essential Oil	55.2*	23	0.97	<0.001
Catnip/Elemol Mixture	93.0*	11	0.83	<0.001
Catnip/Amyris Mixture	82.6*	20	0.94	<0.001
Elemol	63.6*	53	0.97	<0.001
Control	6.8	17	0.19	-

[a] Percentage repellency was determined at 15 minutes.

*Significantly different from control ($\alpha = 0.05$) in LS means comparison.

[c] Avoidance frequency = average of mosquito contact repellency over 3-hour time period.

[d] Contact repellency = 100% of the individuals off treated surface.

Elemol makes up approximately 10% of the Amyris essential oil, along with a collection of other oxygenated sesquiterpenes (eudesmols, valerianol, etc.). Our laboratory has previously reported the mosquito repellent activity of elemol (*21*). When tested for spatial and contact mosquito repellency, elemol showed similar characteristics to its parent essential oil; significant spatial repellency that, on average is lower than catnip essential oil, but with higher levels of contact repellency. The elemol/catnip essential oil mixture provided a combination of highly significant spatial and contact repellencies.

The differences observed in spatial and contact repellency are also highlighted by the relative concentrations of these volatilized compounds inside the repellency bioassay chamber (Table 2). Higher amounts of *Z,E*- and *E,Z*-nepetalactone isomers (ratio in this sample of catnip essential oil was 75:25 *Z,E* / *E,Z*-nepetalctone) distributed quickly inside the repellency chamber, which would be expected of a good spatial repellent. Elemol and DEET, both highly significant contact repellents did not distribute as far, or as quickly as the nepetalactone isomers inside the chamber. Out of the four compounds tested, the lowest level of volatiles collected were in the DEET applications.

Table 2. Volatile collections (in nmol) of *Z,E* and *E,Z*-nepetalactone from catnip essential oil, elemol, and DEET (78.6 μg/cm^2 application rate) in the static-air glass apparatus using solid-phase microextraction with a PDMS fiber.

		Distance Away From Treated Surface			
Volatiles	*Time*	*8 cm*	*23 cm*	*38 cm*	*53 cm*
Z,E-nepetalactone*	15 min.	113	29	3	0
(V.P. = 1.75 mmHg)	30 min.	116	24	12	11
E,Z-nepetalactone*	15 min.	34	10	4	0
(V.P. = 1.75 mmHg)	30 min.	36	9	6	6
Elemol	15 min.	2	2	1	1
(V.P. = 0.24 mmHg)	30 min.	2	2	1	0
DEET	15 min.	1	0	0	0
(V.P. =0.58 mmHg)	30 min.	4	0	0	0

V.P. = vapor pressure (100°C) calculated by ACD Boiling Point software, Version 8.0.
*Isomer measurements made from surfaces treated with catnip essential oil.

Siam-wood essential oil was tested for efficacy in the short-term residual mosquito repellency bioassay. Results for these tests showed good residual spatial and contact repellency (Table 3).

Table 3. Spatial and contact repellency of yellow fever mosquitoes (*Aedes aegypti*) exposed to 78.6 μg/cm^2 application rate of Siam-wood and catnip essential oils and mixtures in the static-air repellency chamber.

	Percentage Repellency over Time				Avoidance	Contact Rep.[b]
Treatment	*1 hr*	*2 hr*	*3 hr*	*6 hr*	*Frequency*[a]	*(P value)*
Catnip Essential Oil	20.3	100% Mortality --------			0.25	0.217
Siam-wood Oil	82.2	92.9	96.3	72	1.00	<0.001
Catnip/Siam-wood Mixture (1:1)	74.1	74.1	100% Mortality		0.83	<0.001
Control	7.4	-14	-18	3.7	0	-

[a] Avoidance frequency = average of mosquito contact repellency over 3-hour time period.
[b] Contact repellency = 100% of the individuals off treated surface.

Some Siam-wood toxicity effects were observed in the repellency screening trials and motivated a house fly LD_{50} toxicity test with the two major components in its essential oil, fokienol and trans-nerolidol (Table 4).

Table 4. House fly 24-hour toxicity to trans-nerolidol and fokienol, two major components in Siam-wood essential oil.

Treatment	LD_{50}	*95% C. I.*
Nerolidol	0.17 µmol/fly	0.14 - 0.21
Fokienol	0.21 µmol/fly	0.12 - 0.34

Amyris (good contact repellent) and catnip (good spatial repellent) essential oils were selected for further testing against active components that are presently used in commercial topical mosquito products. Amyris and catnip essential oils, and p-menthane-3,8-diol were the only three actives to significantly differ in percentage repellency from the control in this study.

Table 5. Spatial and contact repellency tests with yellow fever mosquitoes (*Aedes aegypti*) to surfaces treated with active ingredients (78.6 µg/cm^2 application rate) of commercially available botanical-based repellent candidates and our targeted essential oils in a static-air repellency chamber.

		Average Percentage Repellency				
Product Name	*Active Ingredient*	*15 min.*	*30 min.*	*1 hr.*	Avoidance Frequency[a]	*Contact Rep.*[b] (P value)
OFF® Botanicals (SC Johnson)	p-menthane -3,8-diol	40	80*	78*	1.00	<0.001
BioUD	2-undecanone	41	19	16	0.33	0.093
SCENT OFF TWIST-ONS (ScentOff Corp)	Citronella Oil	52*	44	13	0.75	<0.001
Technical Grade	Catnip Oil	74*	59*	54*	0.66	0.001
Technical Grade	Amyris Oil	27	40	62*	0.83	<0.001
Solvent	Control	-11	-7.4	3.7	0	-

*Significantly different from control ($\alpha = 0.05$) in LS means comparison.

[a] Avoidance frequency = average of mosquito contact repellency over 1-hour time period.

[b] Contact repellency = 100% of the individuals off treated surface.

A small-scale 'barrier' test was used to study brown dog tick repellency. Amyris essential oil was evaluated against an untreated control, and two positive standards DEET and pyrethrum (20%).

The resulting tick climbing activity in the untreated control treatment was 65%. Amyris essential oil and DEET significantly repelled brown dog ticks. Out of 20 ticks that were exposed to Amyris essential oil, only one tick climbed past the Amyris essential oil barrier after repeatedly turning around and climbing down to the arena. No ticks crossed the DEET-treated barriers.

Table 6. Climbing activity of the brown dog tick (*Rhipicephalus sanguineus*) when exposed to barrier-treated surfaces.

Treatment	*Application Rate*	*Percentage Climbing Past Barrier*	*Std. Dev.*
Amyris Essential Oil	1.25 mg/cm^2	5*	11.2
DEET	1.25 mg/cm^2	0*	0
Pyrethrum	1.25 mg/cm^2	40	22.3
Control	-	65	22.3

*Significantly different from control ($\alpha = 0.05$) in LS means comparison.

Conclusions

Plant essential oils are a rich source of sesquiterpenes that can both affect insect behavior and cause mortality. In particular, this study focused on essential oils that contain a select number of closely related sesquiterpenes. Amyris and Siam-wood essential oils were both tested and identified as effective mosquito repellents in a laboratory bioassay. Amyris essential oil was also an effective barrier against brown dog ticks. The majority of these essential oil compositions include oxygenated derivatives of farnesane, eudesmane, eremophilane, and elemane sesquiterpenes. Some of these also are present as primary components of other essential oils (American beautyberry bush, Alaska yellow cedar, etc.) that posses repellent properties. However, interpretation of the sesquiterpene functionality is often times confounded by differences of chirality. One such example is the study of gossypol (+) and (-) enantiomers, found in the cotton plant. These enantiomers have been shown to differ in toxicity to herbivores and pathogens (*22, 23*).

The mosquito laboratory assay in this study allowed for differentiation between contact and spatial repellent activities. High percentage repellency values were observed from mosquitoes exposed to catnip essential oil. The majority of individuals preferred to stay > 1 ft away from the treated surface, representing a significant level of spatial repellency when compared to the control. This observed behavior was not surprising considering the relative concentration of the *Z,E*:*E,Z*-nepetalactone isomers that distributed inside the static-air chamber. Spatial repellency of Amyris essential oil, although lower than catnip, was significantly different from the control treatment and comparable with actives contained in commercial mosquito repellents. Contact repellency, which was measured by cumulative observations of mosquito avoidance of the treated surfaces, was highly significant with Amyris oil. Throughout the 3-hour test period, only one individual came in contact with the treated surface. Similar results of high contact and minimal spatial repellency were seen when testing efficacy of elemol. Relative volatility of elemol, one of the primary components of the Amyris essential oil, was also sampled inside the static-air chamber and did not distribute throughout the chamber as quickly as the nepetalactone isomers. These results show that a chemical's volatility can be an important factor for spatial repellency, affecting the concentration that reaches the insect (*24, 25*). Interestingly, this significant spatial repellency did not always align with effective contact repellency. In the catnip trials there were several mosquitoes that came in contact with treated surfaces and there was no significant difference when compared to the control. These results are consistent with previous studies that have noted the minimal residual effects of catnip essential oil (*21*). This end result is similar to residual effects often observed with many of the first-generation natural repellents. Fradin and Day (*26*) evaluated the protection time of several commercially available repellent formulations, including citronella, peppermint oil, cedar oil, lemongrass oil, and geranium oil. On average, these products provided from 1 to 60 min. of protection whereas DEET formulations scored in a range of 200 to 360 min.

Comparison of catnip and Amyris essential oil shows that volatility isn't the only factor contributing to the repellent activity. Studies that explored the activity of vetiver essential oil found that the individual components' volatility was inversely related to termite repellency (*27*). Based on the characteristic differences in mosquito repellent activity, a mixture containing catnip essential oil (which provided good spatial activity) and the sesquiterpene-rich Amyris essential oil (good contact repellency) was tested. This mixture gave excellent mosquito repellency values via both contact and spatial modes of action. One of the major components in Amyris essential oil, elemol, was also made up in a mixture with catnip essential oil and found effective.

Amyris essential oil was selected for further testing against the brown dog tick. In a climbing arena, individuals that were exposed to an Amyris essential oil barrier would not cross it and frequently avoided contact. These findings were compared with results from a DEET-treated barrier, which successfully prevented ticks from climbing past the chemical barrier. A pyrethrum solution was also tested, but did not significantly prevent ticks from climbing past the barrier.

Siam-wood oil, which contains nerolidol and fokeinol, was also tested for efficacy and evaluated in a mixture with catnip essential oil. Results for these tests showed high levels of both spatial and contact mosquito repellency. Additionally, some mosquito mortality was observed at the concentrations tested inside the static-air chamber. The two major components of Siam-wood were identified as significantly toxic to house flies. To our knowledge, this is the first documented report of insect repellency and toxicological investigation of Siam-wood essential oil.

These findings highlight the potential use of catnip, Amyris, and Siam-wood essential oils for arthropod management. Although the specific repellency mode action of these oils appears to differ in terms of contact and spatial activity, formulated combinations of these did show improvement in a controlled laboratory setting. It is possible that similar mixtures might increase protection efficacy of other natural products.

Acknowledgements

Special thanks to Dianna Wilkening, Theresa Redding, and Ashley Bienemann for technical assistance in the laboratory and the ISU Medical Entomology Laboratory for supplying mosquitoes. Contents of this paper were presented at the 233rd National American Chemical Society Meeting, Chicago, IL, March 25th to 29th, 2007

References

1. Langenheim, J. H.; *J. Chem. Ecol.* **1994**. *20*, 1223-1280.
2. Gershenzon, J.; Dudareva, N. *Nature Chemical Biology*. **2007**. *3*, 408-414.
3. Ahmad, F.; Jantan, I. *Flavour Fragr. J.* **2003**. *18*, 128-130.
4. Watanabe, Y.; Mihara, R.; Mitsunaga, T.; Yoshimura, T. *J. Wood Sci.* **2005**. *51*, 514-519.
5. Wang, S. Y.; Lai, W. C.; Chu, F. H.; Lin, C. T.; Shen, S. Y.; Chang, S. T. *J. Wood Sci.* **2006**. *52*, 522-526.
6. Yatagai, M.; Makihara, H.; Oba, K. *J. Wood Sci.* **2002**. *48*, 51-55.
7. Carroll, J. F.; Cantrell, C. L.; Klun, J. A.; Kramer, M. *Exp. Appl. Acarol.* **2007**. *41*, 215-224.
8. Dietrich, G.; Dolan, M. C.; Peralta-Cruz, J.; Schmidt, J.; Piesman, J.; Eisen, R. J.; Karchesy, J. J. *J. Med. Entomol.* **2006**. *43*, 957-961.
9. Weiss, E. A. Essential Oil Crops. CAB International, New York, NY. 1997. p 504.
10. Jirovetz, L.; Buchbauer, G.; Denkova, Z.; Stoyanova, A.; Murgov, I.; Gearon, V.; Birkbeck, S.; Schmidt, E.; Geissler, M. *Flavour Fragr. J.* **2006**. *21*, 465-468.
11. Howes, M. J. R.; Simmonds, M. S. J.; Kite, G. C. *Journal of Chromatography A*. **2004**. *1028*, 307-312.
12. Miyazawa, M.; Shimamura, H.; Nakamura, S.; Kameoka, H. *J. Agric. Food Chem.* **1996**. *44*, 1647-1650.
13. Schultz, G. E.; Coats, J. R. In *Agrochemical Education Award, Division of Agrochemicals*, 231st American Chemical Society Meeting, Atlanta, GA, March 26-30, **2005**.
14. Schreck, C. E.; McGovern. T. P. *J. Am. Mosq. Control Assoc.* **1989**. *5*, 247-252.
15. WHO. *Report of the WHO Informal Consultation on the Evaluation and Testing of Insecticides*; CTD/WHOPES/IC/96.1; Geneva, Switzerland, **1996**.
16. Amer, A; Mehlhorn, H. *Parasitol. Res.* **2006**. *99*: 478-490.
17. Amer, A.; Mehlhorn, H. *Parasitol. Res.* **2006**. *99*, 473-477.
18. World Wildlife Fund. Annamites trees: Keteleeria evelyniana, *Fokienia hodginsii.* Factsheet. http://www.panda.org/index.cfm (accessed Sept 01 **2007**).
19. Weyerstahl, P.; Marschall, H.; Song, P.T.; Giang P.M. *Flavor Fragr. J.* **1999.** *14*, 409-410.

20. Pavia, D. L.; Lampman, G. M.; Kriz, G. S. Introduction to organic laboratory techniques, 3rd ed. Harcourt Brace College Publishers, Forth Worth, TX. 1988.
21. Schultz, G. E.; Peterson, C.; Coats, J. In Natural Products for Pest Management. Editors, A. M. Rimando, and S. O. Duke. ACS Sym. Ser. 927; American Chemical Society: Washington, DC, 2006. pp. 168 – 181.
22. Stipanovic, R. D.; Puckhaber, L. S.; Bell, A. A.; Percival A. E.; Jacobs, J. *J. Agric. Food Chem.* **2005**. *53*, 6266-6271.
23. Gonzalex-Garza, M. T.; Matlin, S. A.; Mata-Cardenas, B. D.; Said-Fernandez, S. *Arch. Med. Res.* **1992**. *23*, 69-70.
24. Johnson, H. L.; Skinner, W. A.; Maibach, H. I.; Pearson, T. R. *J. Econ. Entomol.* **1967**. *60*, 173-176.
25. Davis, E. E.; Rebert, C. S. *J. Econ. Entomol.* **1976**. *105*, 1058-1061.
26. Fradin, M. S.; Day, J. F. *N. Engl. J. Med.* **2002**. *347*, 13-18.
27. Zhu, B. C.; Henderson, R. G.; Chen, F.; Fei, H.; Laine, R. A. *J. Chem. Ecol.* **2001**. *27*, 1617-1625.

Chapter 3

Structure-activity relationships of naphthalene and 10 related compounds on *Coptotermes formosanus* (Isoptera: Rhinotermitidae)

Sanaa A. Ibrahim[1,2], Gregg Henderson[2*] and Roger A. Laine[2,3]

[1]*Visiting Professor*, **Department of Plant Protection, Faculty of Agriculture, Minia University, Minia, Egypt.**
[2]*Department of Entomology*, **Louisiana State University Agricultural Center, Baton Rouge, LA 70803, USA.**
[3]**Department of Biological Sciences, Louisiana State University, Baton Rouge, LA 70803, USA.**
***To whom correspondence should be addressed**

Naphthalene and ten derivatives were evaluated for initial and residual toxicity, route of penetration and speed of toxic action on *C. formosanus*. In no-choice treated filter paper assays using two colonies, 1'- and 2'-acetonaphthone had the greatest contact toxicity followed by 1- and 2-methoxynaphthalene; toxicity of these chemicals was 7- to 38-fold greater than naphthalene. 2, 7- and 2, 6-diisopropylnaphthalene were 4- to 11-fold less toxic than naphthalene. For all chemicals tested, the colony collected from Lake Charles, LA, was more tolerant than that collected from New Orleans, LA. When termites placed on filter papers treated with the estimated 24 h LC_{90}s, 2'-acetonaphthone followed by 1-methylnaphthalene and 1'-acetonaphthone were the fastest acting toxicants killing 50% of workers after $\leq$ 5 h compared to 16 h with naphthalene. Workers responded faster than soldiers to 1'-acetonaphthone and both responded similarly to 2'-acetonaphthone. At the estimated concentrations for 90% contact mortality, termite mortality via inhalation was not significantly different from the controls in 1'-acetonaphthone, 2'-acetonaphthone and 2-naphthalene methanol treatments.

Naphthalene, 2-isopropylnaphthalene, 1- and 2-methylnaphthalene and 1- and 2-methoxynaphthalene were highly volatile causing 61% to 100% termite mortality via their toxic fumes. In no-choice treated sand assays at 100mg kg^{-1}, 1'-, 2'-acetonaphthone, 1-, 2-methoxynaphthalene and 2-naphthalene methanol were effective toxicants. 1'- and 2'-acetonaphthone maintained their initial toxicity when 1-month residual activity was evaluated. Acetyl substitutions altered the physical and chemical properties of naphthalene moiety to low volatility, more contact toxicity, fast action, and long persistence. This study points to the potential value of 1'- and 2'-acetonaphthone in termite control programs.

Introduction

The Formosan subterranean termite was first described in Formosa (Taiwan) in 1909 and was well established in Louisiana, USA in 1966 (*1*). Compared to other subterranean species, *Coptotermes formosanus* Shiraki is more destructive, more difficult to control, and is responsible for the greatest costs of termite control (*2, 3, 4*). Big colony size and aggressive foraging behavior of the Formosan subterranean termite complicate its control. A mature colony can contain up to 10 million termites, and its foraging area may cover 3577 m^2. Moreover, Formosan termites attack both living trees and structural wood, and can form aerial colonies that do not have a ground contact. As a result, once established it has never been eradicated from an area.[1]

Safe alternatives to synthetic pesticides for termite control are needed because some have been reported to cause air pollution (*5*), contaminate small ponds and poison fish (*6, 7*) and accumulate in body tissues of human (*8*) and other animals.[2] As part of our continuing search for environmentally safe termite control agents, one of the naphthalene derivatives, 2'-acetonaphthone was evaluated on Formosan subterranean termites (*9, 10*) and determined to be a termite toxicant and repellent; affecting tunneling and feeding behaviors at much lower concentrations compared to some other naturally occurring substances such as neem insecticide (*11*); eugenol (*12*); nootkatone (*13*). The findings with 2'-acetonaphthone encouraged us to evaluate more derivatives of naphthalene on the Formosan subterranean termite. Naphthalene derivatives that posses relatively low mammalian toxicity, long stability and low cost were chosen for our current study.

Naphthalene, a bicyclic aromatic hydrocarbon, is known as a toxicant to some insect species (*14*) and is commonly used in houses as a fumigant against cloth moths and carpet beetles (*15*). It is used as an external medication to control lice on livestock and poultry (*16*). Naphthalene has low mammalian toxicity, with oral LD_{50}s of 1200 mg kg^{-1} (guinea pigs), 553 mg kg^{-1} (mice) and

[1] http://www.aces.edu/department/ipm/formoterm.htm
[2] http://ipm.ncsu.edu/wildlife/cotton_wildlife.html

490 mg kg^{-1} (rats) (*17*). In addition, the mutagenic effects of naphthalene *in vitro* and *in vivo* were negative (*18, 19, 20*). However, naphthalene has been reported to cause hemolytic anemia in the people from Mediterranean countries after long time exposure to very high concentrations (*15*). Reduced concentrations of glutathione in rat and mice neonates are also a reported side effect of a long-term naphthalene exposure (*16, 21*).

To our knowledge there are no publications on the effect of naphthalene and the derivatives we tested on termites, except that naphthalene was surprisingly found in termite carton nests at 50.56-214.6 μg kg^{-1} and is believed to constitute a unique chemical defense strategy against natural enemies of the Formosan subterranean termite (*22*). In addition, Formosan subterranean termites have been found to follow trails of naphthalene (*23*) and its derivative, 2-naphthalene methanol (*24*) and may be useful as termite bait additives. A derivative of naphthalene, N,N-naphthalolylhydroxylamine, was evaluated for its efficiency as a fungicide and a termiticide (25). Also copper naphthenate has been proven to be effective in preventing the consumption of wood by the aggressive Formosan termite in field and laboratory tests.[3]

Naphthalene derivatives selected for our study are naturally occurring chemicals found in petroleum oil (*15, 26*). 2-Methylnaphthalene was identified as a volatile constituent of dried legumes at concentrations that ranged 2.8 to 49.2 ppb (*27*). 1'- and 2'-acetonaphthone were identified in corn bud essential oil (*28*) and both are listed by Fisher Scientific CANADA[4] and Chemical Land[5] as major constituents in the fragrances of perfumes and household products. 2, 6-Diisopropylnaphthalene is used to inhibit sprouting in potatoes held in the storage (*29*) and also used to prepare Naproxen [2-(6-methoxy-2-naphthyl) propionic acid], which is used as a non-steroidal anti-inflammatory drug (*30*). Di-isopropylnaphthalene (D-IPN) is used as a solvent for ink and has been identified in samples of food packaging materials made from recycled board and in some samples of food (*31*).

We compared the performance of these derivatives on the Formosan subterranean termite since they have low mammalian toxicity (*26*) compared to commonly used termiticides; (*32*) their oral rat LD_{50}s ranged 599 to >5000 mg kg^{-1}.[6,7] Also they contain no chemical groups, which would be structurally altering for potential mutagenicity (naphthalene, 1-methylnaphthalene and 2-methylnaphthalene (*26*); 2, 6-Diisopropylnaphthalene (*29*); 2-isopropylnaphthalene and 2'-acetonaphthone (*33*); di-isopropylnaphthalene (*34*). In addition, they are relatively stable and inexpensive. The current study includes toxicity, speed of toxic action, route of penetration and longevity of these chemicals in relation to substitutions on the fused-ring system of naphthalene.

---

[3] http://www.merichem.com/Copper/CuNapRpt11.htm
[4] http://www.fishersci.ca/msds2.nsf/EView1/11647/$file/msds-11647.html
[5] http://www.chemicalland21.com/arokorhi/specialtychem/perchem/2'-ACETONAPHTHONE.htm
[6] http://www.sigmaaldrich.com/
[7] http://ptcl.chem.ox.ac.uk/MSDS/ME/2-methoxynaphthalene.html

Materials and Methods

Termites and chemicals

Termites from two colonies were used in this study. Termites from colony A were collected from an island along the Calcasieu River in Lake Charles, LA on January 6, 2003. Termites from colony B were collected from Brechtel Park, New Orleans, LA on January 15, 2003. Naphthalene (98% purity); 1'-acetonaphthone (98%); 2'-acetonaphthone (99%); 1-methoxynaphthalene (98%); 2-methoxynaphthalene (99%), 2-naphthalene methanol (98%); 1-methylnaphthalene (95%); and 2-methylnaphthalene (97%) were purchased from Aldrich Chem. Co. Inc., Milwaukee, WI. 2-Isopropylnaphthalene (95%); 2, 6-diisopropylnaphthalene (99%); and 2, 7-diisopropylnaphthalene (95%) were purchased from TCI American, Portland, OR (see Figure 1). Absolute ethanol (Ethyl alcohol USP, absolute-200 proof, Aaper Alcohol and Chemical Co. DSP-KY-417, Shelbyville, KY) was used as a solvent for all chemicals except naphthalene and 2, 6-diisopropylnaphthalene, which were dissolved in n-hexane (J T Baker Chemical Co, Phillipsburg, NJ).

Acute toxicity

For each chemical treatment, filter papers (Whatman # 2, 55 mm diameter, Whatman International Ltd, Maidstone, England) were fitted in plastic Petri dishes (6cm diameter by 1.5 cm high) and coated with the tested concentrations dissolved in 250µl solvent. Filter papers treated with solvent only served as the control. Five replicates were performed for each treatment. Containers were left for 4 h uncovered at ambient conditions and then 10 workers were placed in each container after the filter paper was moistened with 250 µl double distilled water (DDH_2O). Petri dishes were covered and incubated (26.4 °C, 59% RH, darkness) for 24 h before worker mortality was recorded. For each chemical and colony the 24-h mortality data were corrected by using Abbott's transformation (*35*), and then probit analysis results were established (*36*).

Speed of toxic action

For each chemical evaluated, six plastic containers (5.5 cm diameter by 3.7 cm high) each were provided with a Whatman # 2-filter paper. In three containers, filter papers were coated with the appropriate amount of the chemical in 250 µl solvent to yield a concentration equivalent to the 24 h LC_{90} as µg cm^{-2}. Filter papers in the other three containers were coated with 250 µl solvent and served as a control. After drying at ambient conditions (4 h), filter papers in the six containers were moistened with 250 µl DDH_2O and provided with 100 workers from colony A. The containers were covered with their lids and an opaque black sheet and kept at laboratory conditions for short-term observations. Ten readings at one-hour intervals followed by seven readings every two hours were made to record mortality.

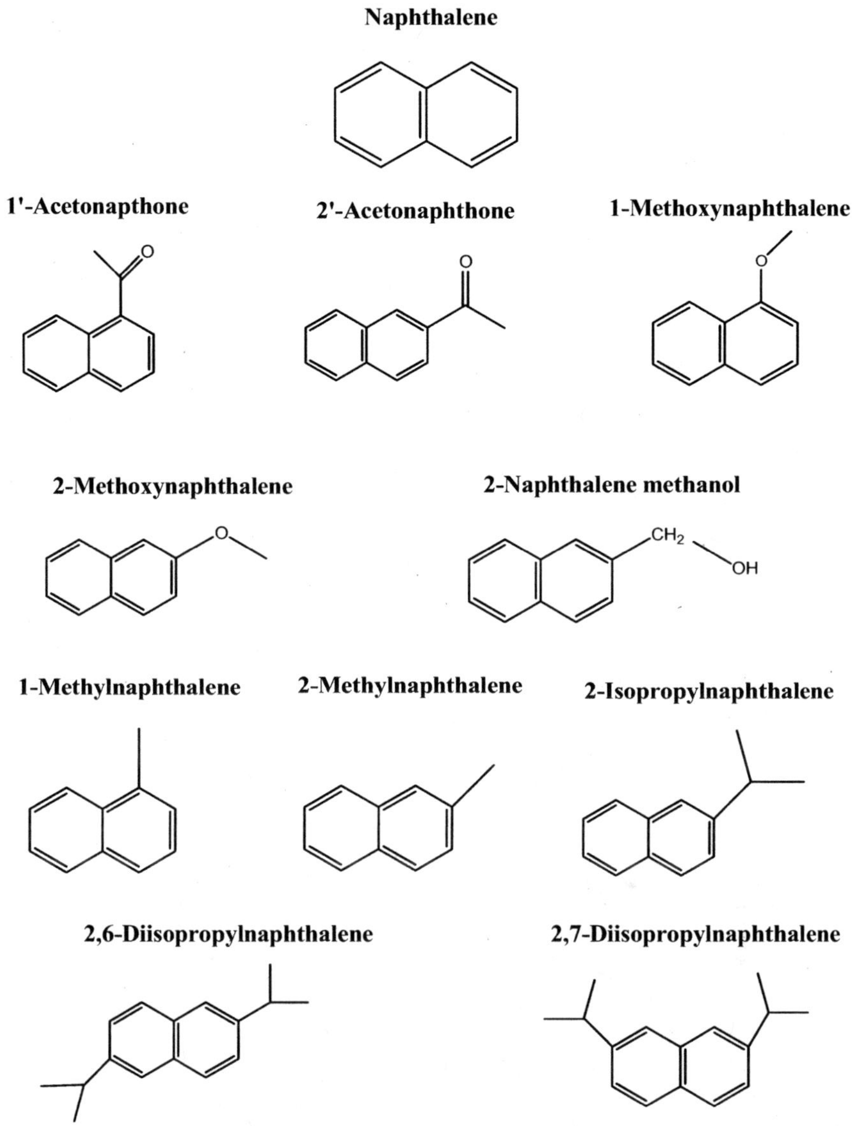

Figure 1. Chemical structure of naphthalene and 10 derivatives.

Based on the data obtained with the previous experiment, another experiment was conducted using the most effective chemicals, 1'-acetonaphthone, 2'-acetonaphthone, 1- and 2-methoxynaphthalene on workers and soldiers from colony B. The previous technique was used except: 1) a diagnostic concentration (20 μg cm^{-2}) was used because the number of soldiers was not enough to establish the probit analysis results; 2) eight replicates of ten soldiers or ten workers each was used for each chemical assay and control; and 3) observations were recorded for 10 h at one hour intervals and an additional reading was recorded after 24 h. For each chemical, mortality at each time interval was corrected with control mortality using Abbott's formula (*35*) and probit analysis results were calculated (*36*).

Route of exposure

To establish whether toxicants are transmitted by physical contact or via inhalation, a technique developed by Delgarde and Rouland-Lefevre (*37*) was used with some modifications. Worker mortality in the untreated enclosures away from any physical contact with the chemicals tested was used as an indicator of the inhalation route of penetration. We chose the concentrations that induced the same toxicity response via physical contact (24 h LC_{90}s) to avoid effects that may be related to the variation in the toxicity of the chemicals. For each chemical and colony tested, Whatman # 2-filter papers were fitted on the bottom of 48 plastic containers (5.5 cm diameter by 3.7 cm high). Twenty-four containers were marked "treated" and filter paper in each container was coated with 250 µl of the chemical solution adjusted to have the 24 h LC_{90} as $\mu g\ cm^{-2}$. Filter papers in another 24 containers were left untreated. The 48 containers of the control were handled the same except that filter paper in each of the 24 containers that marked "treated" received 250 µl solvent only. Treated containers with either chemical solution or solvent were kept 4 h uncovered at ambient conditions for solvent evaporation. Filter papers in all containers were wetted with 250 µl DDH_2O before introducing 20 workers. For either control or chemical treatment, the 48 containers were divided into 6 sub-groups each was consisting of 4 containers with treated filter papers and 4 with untreated filter papers, were housed together uncovered in a large plastic container (20 cm diameter by 7.8 cm high). The large plastic container was covered with its lid and incubated (26.4 °C, 59% RH, darkness) for 24 h. This technique allowed for the dispersion of the chemical vapors from the treated into the untreated containers preventing direct contact between termites in the untreated enclosures and the chemical. Mortality in treated and untreated enclosures was recorded and corrected with the corresponding mortality in the controls (*35*). For each chemical, among the two colonies (A & B) and the two routes of exposure (physical contact & inhalation), mortality percentages were analyzed using SAS GLM procedure followed by Tukey's Studentized Range (HSD) Test (*38*).

A second experiment was conducted to confirm the previous finding with 2-naphthalene methanol, 1'- and 2'-acetonaphthone. The same technique described above was used except using termites from different colony (colony B) and mortality in treated and untreated enclosures was recorded for six days at 1-day intervals.

Initial and residual activity

Two hundred gram sand (fine blasting sand # 4, Cement Products Inc., Baton Rouge, LA) was held in a plastic container (20 cm diameter by 7.8 cm high) and mixed with 25 ml from a stock solution [800 mg (AI) $litre^{-1}$ ethanol] for each chemical, to yield a final concentration of 100 mg kg^{-1} sand. The same amount of sand receiving the same volume of ethanol served as a control. Containers having chemical- and ethanol-treated sand were kept uncovered overnight at ambient conditions for ethanol evaporation. To evaluate the initial toxicity, 100 g of treated sand was mixed with 10 ml DDH_2O. Blaine Test Disc (S & S # 597, 12.7 mm diameter, Keene, NH) was centered in a plastic Petri dish (6cm diameter by 1.5 cm high). Eleven grams of wetted sand was placed in each Petri dish, leveled and packed and then 20 workers from colony A were placed on the surface of the sand. Each treatment was replicated 10 times. Petri dishes were covered with their lids and incubated (26.4°C, 59% RH, darkness) for 11 days (checked daily to observe mortality and suitable moisture). On day 11, the bottoms of the containers were scanned to obtain an image of tunnels and the filter paper disc. The number of living workers in each replicate was then counted. Squared areas of consumed filter paper and the tunnels constructed were measured from the printed images. For studying the residual activity of the tested chemicals, the other half of each batch of sand was kept at 26.4 °C, 59% RH and darkness for 1-month in a glass jar. Stored sand was evaluated on workers from the same colony using exactly the same technique as described above. Among treatments, mean percentages of mortality, mean tunnel areas and mean food consumption were subjected to SAS GLM procedure followed by Tukey's (HSD) Test (*38*).

Results

Acute contact toxicity

Of the 11 chemicals tested, 1'- and 2'-acetonaphthone had the greatest acute toxicity after 24h exposure to treated filter paper (Table I). 1'- and 2'-acetonaphthone exhibited similar toxicity and were significantly more toxic than 1- and 2-methoxynaphthalene. 1- and 2-methoxynaphthalene were more toxic than the rest of the chemicals including naphthalene. Toxicity of 1'-, 2'-acetonaphthone, 1- and 2-methoxynaphthalene (based on the LC_{50}s, Table I) was 7 to 38-fold (for colony A) and 14 to 22-fold (for colony B) greater than that of naphthalene. Acute toxicity of 2-naphthalene methanol (colony B), 1-methylnaphthalene (colony B) and 2-methylnaphthalene (colonies A & B) was not significantly different from naphthalene (based on the overlap of 95% CL of the LC_{50}s, Table I). 1-Methylnaphthalene (colony A), 2-isopropylnaphthalene (colony A), 2-naphthalene methanol (colony A), 2, 6- and 2, 7-diisopropylnaphthalene (colonies A & B) were significantly less toxic than naphthalene. For all of the tested chemicals, colony A (Lake Charles) was significantly more tolerant (based on the non overlap of 95% confidence limits of the LC_{50}s) than colony B (New Orleans). Of the 11 chemicals tested, 2, 6-diisopropylnaphthalene was the least toxic chemical; moreover, workers from colony A did not respond to any of the tested concentrations of 2, 6-diisopropylnaphthalene up to 2000 μg cm^{-2}.

Speed of toxic action

2'-Acetonaphthone followed by 1-methylnaphthalene and 1'-acetonaphthone were the fastest acting toxicants (based on the LT_{50}s, Table II). 1- and 2-methoxynaphthalene were relatively slow acting compared to 1-methylnaphthalene and 1'-and 2'-acetonaphthones; but they acted similarly and significantly faster than naphthalene, 2-naphthalene methanol, 2, 6-diisopropylnaphthalene and 2, 7-diisopropylnaphthalene. 2-Isopropylnaphthalene and 2-methylnaphthalene had statistically the same speed of action of 1-methoxynaphthalene. 2-Naphthalene methanol and 2, 6-diisopropylnaphthalene took longer time to induce similar toxicity response.

The potency and fast action of 1'-, 2'-acetonaphthone, 1- and 2-methoxynaphthalene encouraged us to re-evaluate them at 20 μg cm^{-2} on workers and soldiers from colony B (Table III). 1'- and 2'-acetonaphthone were similarly acting on workers and both resulted in 100% worker mortality after 5h (LT_{50} was 2.58 h and 2.37 h; respectively). However, when they were evaluated on soldiers, 1'-acetonaphthone was significantly slower acting than 2'-acetonaphthone; inducing 100% mortality after 10 h and 7 h, respectively (LT_{50} was 5.62 h and 3.30 h, respectively). Workers responded significantly faster than soldiers to 1'-acetonaphthone; however, both responded similarly to the toxic action of 2'-acetonaphthone. In general, 1'- and 2'-acetonaphthones were about 3-fold faster acting on workers than 1- and 2-methoxynaphthalene. Although 1-and 2-methoxynaphthalene induced 77% and 100% worker mortality, respectively after 10 h; however, their speed of toxic action was not significantly different (Table III). Soldiers did not show signs of toxicity after 10 h exposure to either of these two tested chemicals; however after 24 h exposure, all termite workers and soldiers were dead.

Table I. Contact toxicity of naphthalene and 10 derivatives on Formosan subterranean termite workers

Chemical	Colony, n[a]	Slope ± SE	χ^2, *df*, *p*	LC_{50}[c] (95% CL)	LC_{90}[c] (95% CL)
Naphthalene	A, 1050	4.1 ± 0.4	21.2,18, 0.3	264.6 (232.4-296.4)c	547.9 (447.8-815.4)d
	B, 900	3.0 ± 0.2	18.8, 15, 0.2	100.9 (83.8-116.9)d	274.1 (223.5-376.7)e
1'-Acetonaphthone	A, 800	4.1 ± 0.2	19.5, 13, 0.1	6.9 (6.3-7.6)g	14.2 (12.4-16.9)h
	B, 500	7.2 ± 1.2	9.4, 7, 0.2	5.0 (2.8-5.8)h	7.6 (6.8-10.9)ij
2'-Acetonaphthone	A, 800	3.5 ± 0.2	14.5, 13, 0.3	7.9 (7.1-8.8)g	18.4 (15.7-22.9)h
	B, 600	6.7 ± 0.6	2.2, 9, 1.0	4.6 (4.2-4.9)h	7.1 (6.7-7.6)j
1-Methoxynaphthalene	A, 750	6.5 ± 0.5	20.1, 12, 0.1	36.1 (32.8-39.4)f	56.8 (50.1-70.9)g
	B, 600	3.8 ± 0.3	14.8, 9, 0.1	7.4 (6.5-9.2)g	16.0 (12.6-25.2)h
2-Methoxynaphthalene	A, 800	3.3 ± 0.2	11.2, 13, 0.6	29.6 (25.8-33.9)f	73.2 (59.2-101.7)g
	B, 650	5.2 ± 0.4	15.9, 10, 0.1	7.2 (5.9-8.3)g	12.7 (10.7-17.3)hi
2-Naphthalene methanol	A, 900	1.1 ± 0.1	8.1, 15, 0.9	991.1 (772.4-1382.3)a	14352.3 (7439.3-39284.9)a
	B, 850	2.2 ± 0.1	18.3, 14, 0.2	110.3 (91.4-129.1)d	421.2 (356.0-518.3)de
1-Methylnaphthalene	A, 750	3.2 ± 0.3	13.0, 12, 0.4	451.3 (350.5-541.6)b	1144.4 (867.5-2115.8)c
	B, 500	1.9 ± 0.1	5.7, 7, 0.6	74.4 (42.0-107.6)de	352.6 (227.9-810.5)de
2-Methylnaphthalene	A, 650	4.9 ± 0.4	15.4, 10, 0.1	277.3 (235.3-312.3)c	503.9 (430.5-674.5)d
	B, 450	2.4 ± 0.2	11.9, 6, 0.1	79.6 (33.4-121.1)def	269.3 (173.9-746.5)def
2-Isopropylnaphthalene	A, 700	2.0 ± 0.3	7.9, 11, 0.7	1108.8 (894.6-1583.8)a	4844.0 (2874.4-12468.9)ab
	B, 500	2.42± 0.2	8.6, 7, 0.3	47.4 (36.9-57.1)ef	159.2 (132.2-199.9)f
2, 6-				ND	ND
Diisopropylnaphthalene	A	ND[b]	ND	1153.4 (1024.6-1255.3)a	1933.3 (1708.0-52426.8)c
	B, 800	5.7 ± 0.6	12.3, 13, 0.5		
2, 7-				1139.7 (1022.7-1286.2)a	3490.6 (2821.6-4642.0)b
Diisopropylnaphthalene	A, 750	2.6 ± 0.2	18.2, 12, 0.1	429.9 (358.5-501.9)b	1000.8 (832.0-1294.9)c
	B, 650	3.5 ± 0.2	23.4, 10, 0.0		

[a] Number of workers tested. [b] Not determined.

[c] The LC_{50}s and LC_{90}s are the lethal concentrations (μg cm^{-2}) for 50 and 90% of termite workers. For each column, LC_{50}s or LC_{90}s followed by the same letters are not significantly different (based on the overlap of 95% confidence. Controll mortality was 6.0 ± 1.9 (for colony A) and 0.0 (for colony A).

Table II. Logit time-probit analysis results of naphthalene and 10 derivatives on Formosan subterranean workers[a]

Chemical	Slope[b] ± SE	χ^2, *df*, *p*	LT_{50} (95% CL)[d]	LT_{90} (95% CL)[d]
Naphthalene	6.7 ± 0.3	15.3, 12, 0.2	16.1 (15.0-17.3)b	25.0 (22.4-29.1)cd
1'-Acetonaphthone	5.0 ± 0.3	17.4, 11, 0.1	4.7 (4.3-5.1)f	8.4 (7.6-9.5)f
2'-Acetonaphthone	5.7 ± 0.4	5.1, 5, 0.4	3.1 (2.8-3.3)g	5.1 (4.7-5.7)g
1-Methoxynaphthalene	4.5 ± 0.2	7.5, 15, 0.9	8.7 (5.6-12.4)cde	16.8 (11.9-47.7)cde
2-Methoxynaphthalene	6.7 ± 0.3	10.8, 15, 0.8	9.5 (9.2-9.8)d	14.7 (14.0-15.5)e
2-Naphthalene methanol	7.0 ± 2.0[c]	2.1, 15, 1.0	39.0 (31.1-87.8)a	59.5 (40.8-236.5)ab
1-Methylnaphthalene	3.0 ± 0.2	17.6, 6, 0.0	3.5 (2.4-4.6)fg	9.4 (6.5-22.9)def
2-Methylnaphthalene	2.2 ± 0.1	17.8, 15, 0.3	7.1 (6.5-7.7)e	26.7 (23.2-31.8)c
2-Isopropylnaphthalene	4.1 ± 0.2	16.6, 10, 0.1	11.1 (10.3-11.9)c	22.9 (20.6-26.4)cd
2, 6-Diisopropylnaphthalene	5.2 ± 0.8[c]	6.9, 15, 1.0	34 (29.7-43.0)a	59.7 (46.3-93.9)a
2, 7-Diisopropylnaphthalene	3.5 ± 0.2	9.9, 15, 0.8	16.7 (15.8-17.7)b	38.4 (34.2-44.1)b

[a]Termite workers from colony A were used.
[b]Number of insects on which each probit analysis based was 600.
[c]The estimated LC_{90}s with colony B were used to evaluate the speed of toxic action on colony A.
[d]LT_{50}s and LT_{90}s expressed as time in hours required to kill 50 and 90% of termite workers. For each column, LT_{50}s and LT_{90}s followed by the same letters are not significantly different (based on the overlap of 95%. The 24 h control mortality was 3.3 ± 1.2 (for colony A) and 8.6 ± 2.5 (for colonyB).

Table III. Logit time-probit analysis results when Formosan subterranean termite workers and soldiers (colony B) were exposed to filter paper treated with the tested chemicals at 20µg cm^{-2}

Chemical	Termite group[a]	Slope ± SE	χ^2, *df*, *P*	LT_{50} (95% CL)[b]	LT_{90} (95% CL)[b]
1'-Acetonaphthone	Worker	6.7 ± 0.6	6.6, 3, 0.1	2.6 (2.2-3.7)b	3.9 (3.1-4.9)b
	Soldier	4.4 ± 0.3	11.1, 8, 0.2	5.6 (5.3-6.0)a	10.5 (9.5-12.0)a
2'-Acetonaphthone	Worker	5.4 ± 0.4	5.6, 3, 0.1	2.4 (1.5-3.0)b	4.0 (3.1-4.6)b
	Soldier	6.0 ± 0.4	2.8, 5, 0.7	3.3 (2.7-3.9)b	5.2 (4.4-7.2)b
1-Methoxynaphthalene	Worker	4.7 ± 0.4	12.1, 8, 0.2	6.6 (6.0-7.4)a	11.9 (9.8-16.7)a
	Soldier	ND[b]	ND	> 10	> 10
2-Methoxynaphthalene	Worker	8.3 ± 0.5	7.1, 8, 0.5	6.4 (5.5-7.6)a	9.0 (7.7-12.9)a
	Soldier	ND[b]	ND	ND	ND

[a]n = 80.

[b]LT_{50}s and LT_{90}s expressed as time in hours required to kill 50 and 90% of termite workers.

[c]Soldier mortality was not significant during the first 10h observations, however 100% mortality was observed after 24h exposure.

For each column, LT_{50}s and LT_{90}s followed by the same letters are not significantly different (based on the overlap of 95% confidence limits). the 24 hr control mortality was 0.0 (for workers) and 7.5 ± 2.3 (for soldiers).

Route of exposure

With the exception of 2, 6-diisopropylnaphthalene, the estimated concentrations for killing 90% of termite workers via physical contact resulted in 24 h contact mortality ranged 59 to 99% (colony A) and 79 to 100% (colony B) that were not significantly different among chemical treatments (Table IV). However, inhalation mortality was significantly varied between treatments. 1'- and 2'-acetonaphthone (for the two colonies) followed by 2-naphthalene methanol and 2, 7-diisopropylnaphthalene (for one colony) induced inhalation mortality was not significantly different from the control; however was significantly different from the corresponding mortality via physical contact (Table IV). In the assays of the two colonies, naphthalene, 1- and 2-methylnaphthalene, 1- and 2-methoxynaphthalene and 2-isopropylnaphthalene were highly volatile. In each of the 6 treatments, mortality via physical contact and via inhalation was not significantly different; moreover, both were significantly different from the controls (Table IV). Inhalation mortality with workers from colony B in 2, 7-diisopropylnaphthalene treatment was negligible; however, 58% inhalation mortality was achieved with workers from colony A (Table IV). In 2, 6-diisopropylnaphthalene treatment, no remarkable mortality was achieved in either treated or untreated enclosures.

Repeating the experiment with the low-volatile chemicals (1'- and 2'-acetonaphthones and 2-naphthalene methanol) for a six-day observation period (data not shown in a table) revealed that mortality of workers (from colony B) via physical contact with 1'-acetonaphthone was 75% and 92.5% in the first two successive days compared to 0 and 2.5 % worker mortality in the control. Cumulative inhalation mortality in the untreated enclosure of the 1'-acetonaphthone treatment did not exceed 12.5% on day 6 compared to 17.5% in the control. 2'-Acetonaphthone induced 100% mortality after 24 h when termite workers were in physical contact with the 24-h LC_{90}-treated filter paper. However, inhalation mortality in the untreated enclosure was only 32.5% compared with 17.5% mortality in the control. In the enclosures, which had, filter paper treated with 2-naphthalene methanol, mortality via physical contact was 12.5, 75, 86.25 and 100% in the first four successive days. The corresponding mortality via inhalation was 11%, 19%, 28% and 44% compared to 0%, 6%, 8% and 9% in the control.

Initial activity

1'-Acetonaphthone, 2'-acetonaphthone, 1-methoxynaphthalene, 2-methoxynaphthalene and 2-naphthalene methanol were the only effective toxicants, resulting in 98.5% to 100% worker mortality compared to 12% in the control on day 11 (Table V). Complete mortality was recorded at day 2 for 1'-acetonaphthone, 2'-acetonaphthone and at day 3 in 2-methoxynaphthalene. Similar toxic effect required 5 days in 1-methoxynaphthalene and >11 days in 2-naphthalene methanol. Food consumption and tunnel construction were negligible in these treatments after 11 days exposure. We also observed that the abdomens of all dead workers changed to dark blue in 1-methoxynaphthalene treatment. Mortality of termite workers after 11 days exposure to naphthalene and the other tested derivatives was not significantly different from the control (Table V). Tunnel areas were significantly reduced in all chemical treatments except for 1-methylnaphthalene. Also feeding activity was significantly reduced in all treatments except 1-methylnaphthalene, 2-methylnaphthalene and 2, 6-diisopropylnaphthalene. In 1-methylnaphthalene treatments termites tunneled as long as control, however their feeding activity was significantly greater than the control (Table V).

Residual activity

1'- and 2'-acetonaphthone maintained their initial efficiency, killing 99.5% and 100% of the termites, respectively. Moreover, all termites died within three days in 2'-acetonaphthone and consequently complete inhibition of food consumption and tunneling activity was observed (Table V). 1-Methoxynaphthalene, 2-methoxynaphthalene and 2-naphthalene methanol lost most of their initial toxic effects when their residual activity was assayed; however, food consumption and tunnels constructed in the three treatments in addition to 2-isopropylnaphthalene; 2, 6- and 2, 7-diisopropylnaphthalene treatments were significantly reduced compared to the control. Compared to the control, tunnels constructed in the treatments of 2-isopropylnaphthalene and 2, 7-diisopropylnaphthalene were significantly shorter, however, no significant effect on food consumption was achieved (Table V).

Discussion

Naphthalene and its derivatives are dicyclic aromatic hydrocarbons that exhibit toxic properties, which appear to be a function of their fused- ring system (*39*). We found that the double-ring system plays a minor role to toxicity; however, substitutions on the naphthalene moiety significantly altered the toxicity, speed of action, route of penetration, volatility and consequently the residual activity. Of the 10 naphthalene derivatives tested, it was evident that an acetyl group attached to naphthalene in either the 1- or 2-position significantly improved the toxicity and the speed of toxic action. At the same time, this modification altered the route of penetration from inhalation to contact entry and consequently increased the persistence compared to the rest of tested naphthalene derivatives. In a previous study with 2'-acetonaphthone (*9*) termites placed on 40 $\mu g\ cm^{-2}$ treated filter paper died within 6-8 h; however survival of termites exposed in two-choice assays to treated filter paper with the same concentration were not affected for up to 15 days. Greater and faster contact toxicity together with a relatively higher persistence of 1'- and 2'-acetonaphthone were accompanied by lower inhalation toxicity. It has been previously reported that toxicity and lipophilicity of naphthalene derivatives increased and volatility decreased as the alkylation increased (*26, 40*). 1'- and 2'-acetonaphthone are not alkyl derivatives of naphthalene, however they were more toxic and less volatile than naphthalene.

The methoxy group attached to naphthalene in either the 1- or 2-position significantly improved the initial toxicity and the speed of action; however, this modification maintained the high volatility, which allowed the chemical when applied to sand to lose most of toxicity within 1 month. Toxicity was significantly diminished when two isopropyl groups were attached to naphthalene in 2, 6 or 2, 7 positions.

To our knowledge this is the first study regarding the structure-activity relationship of naphthalene and its derivatives on any insect species. However, derivatives of the allylamine antimycotic terbinafine with varied substitution at the naphthalene ring system have been evaluated for their antifungal activity (*41*). They found that substitutions that increase lipophilicity were much more important for toxicity than the electronic density distribution and /or steric requirements. In our study we found that acetyl substitution increased the toxicity and speed of action when compared to naphthalene itself probably through increase in lipophilicity and reduced volatility.

Conclusions

For any insecticide, alterations in the electrophilic properties (*42, 43, 44, 45*), the hydrophobicity (*46*), the flexibility and steric changes (*47*) can affect the affinity of an insecticide to its site of action. In addition, substitutions on the original molecule may affect the cuticular penetration and metabolic degradation (*42, 48*). Changes in the molecule structure may also result in kinetic changes in the function of receptors (*49*). Substitutions that resulted in low volatility and high lipophilicity were associated with the greatest toxicity, the fastest action, and the longest persistence in a group of naphthalene derivatives tested against termites. 1'- and 2'-acetonaphthones when compared to other tested naphthalene derivatives may increase their selectivity toward insects. Non-volatile chemicals are more valuable in pest control for indoor safety and outdoor persistence. The low volatility of 1'- and 2'-acetonaphthone allows them to persist longer under field conditions than volatile chemicals. Both maintained their initial activity in treated sand when 1-month longevity was considered. High toxicity to insects is not always associated with high mammalian toxicity (*50*), even if the mode of action is the same for both. However, differences can be due to the route of entry into the tissue. Lipophilic insecticides are more selective than strongly volatile chemicals because mammals are mostly exposed to chemicals through inhalation, and percutaneous penetration is negligible. This study points to the potential value of 1'- and 2'-acetonaphthone in termite control programs.

Table IV. Percentages of corrected mortality [a] of Formosan subterranean termites exposed to the estimated LC_{90}s of naphthalene and its homologues through physical contact or/and inhalation

	and inhalation		and inhalation		3, 20*; *P*
Control	0.0 ± 0.0 (B)	0.0 ± 0.0 (D)	0.0 ± 0.0 (B)	0.0 ± 0.0 (B)	---
Naphthalene	81.8 ± 15.1a (A)	80.4 ± 15.2a (AB)	100.0 ± 0.0a (A)	100.0 ± 0.0a (A)	1.08; 41.513; 0.3789
1'-Acetonaphthone	59.0 ± 10.6b (A)	13.8 ± 6.0c (CD)	93.5 ± 4.0a (A)	0.0 ± 0.0 c (B)	40.65; 26.529; < 0.0001
2'-Acetonaphthone	71.2 ± 13.6a (A)	32.8 ± 11.1b (BCD)	100.0 ± 0.0a (A)	9.7 ± 2.8b (B)	21.18; 34.481; < 0.0001
1-Methoxynaphthalene	59.4 ± 17.5a (A)	60.9 ± 17.9a (ABC)	82.9 ± 11.1a (A)	67.8 ± 17.9a (A)	0.45; 63.474; 0.7203
2-Methoxynaphthalene	88.7 ± 8.4a (A)	84.2 ± 11.7a (AB)	100.0 ± 0.0a (A)	86.8 ± 8.9a (A)	0.71; 32.775; 0.5571
2-Naphthalene methanol	ND	ND	78.8 ± 14.2a (A)	25.8 ± 5.0b (B)	12.85;32.896; < 0.0050
1-Methylnaphthalene	99.4 ± 0.6a (A)	100.0± 0.0a (A)	100.0 ± 0.0a (A)	100.0 ± 0.0a (A)	1.0; 1.237; 0.4133
2-Methylnaphthalene	76.2 ± 16.2a (A)	75.0± 17.0a (AB)	100.0 ± 0.0a (A)	100.0 ± 0.0a (A)	1.51; 45.463; 0.2426
2-Isopropylnaphthalene (beta-)	88.5 ± 6.5a (A)	94.8 ± 4.8a (A)	83.0 ± 13.3a (A)	71.0 ± 15.8a (A)	0.87; 42.987; 0.4742
2, 6-Diisopropylnaphthalene	ND	ND	29.4 ± 13.0a (B)	23.5 ± 13.3a (B)	0.09; 21.961; 0.7654
2, 7-Diisopropylnaphthalene	65.3 ± 17.4a (A)	58.4 ± 16.6a (ABC)	87.4 ± 10.5a (A)	3.4 ± 2.3b (B)	8.37; 37.652; 0.0008
F; MSD; df; P	5.19; 56.806 9, 50; < 0.0001	8.45; 54.927; 9, 50; < 0.0001	16.08; 38.14; 11, 60; < 0.0001	26.11; 39.973 11, 60; < 0.0001	---

[a]Data expressed as mean ± SE. For each column, means with the same capital letters are not significantly different (P > 0.05). For each row, means with the same lowercase letters are not significantly different (P > 0.05).

* df = 1, 10 in 2-naphthalene methanol and 2, 6-diisopropylnaphthalene treatments

Table V. Mortality percentages, tunnel areas and filter paper consumption[a] of Formosan subterranean workers measured on day 11 in response to 100 mg kg^{-1} treated sand in initial and residual assays

Treatment	% Mortality	Tunnel area (cm^2)	Consumption (mm^2)	% Mortality	Tunnel area (cm^2)	Consumption (mm^2)
Control	11.5 ± 1.1b	8.5 ± 2.7a	8.1 ± 2.3b	12.5 ± 1.3c	8.2 ± 0.9ab	31.0 ± 8.6a
Naphthalene	20.0 ± 4.5b	3.0 ± 1.4bc	0.9 ± 0.5c	23.0 ± 5.1bc	5.2± 0.7bcd	9.5 ± 5.7abc
1'-Acetonaphthone	100.0 ±0.0a	0.0± 0.0c	0.0± 0.0c	99.5 ± 0.5a	0.1 ± 0.0f	0.0± 0.0c
!'-Acetonaphthone	100.0 ±0.0a	0.0± 0.0c	0.0± 0.0c	100.0 ±0.0a	0.0± 0.0f	0.0± 0.0c
1-Methoxynaphthalene	100.0 ±0.0a	0.4± 0.1c	0.0± 0.0c	33.0 ± 5.1bc	0.2 ± 0.0f	1.0 ± 0.7c
2-Methoxynaphthalene	100.0 ±0.0a	0.7± 0.5c	0.0± 0.0c	44.5 ± 6.1b	0.0± 0.0f	3.0 ± 0.8bc
2-Naphthalene methanol	98.5 ± 0.8a	0.5 ± 0.3c	0.0 ± 0.0c	33.0 ± 5.2bc	0.6 ± 0.0ef	1.0 ± 0.7c
1-Methylnaphthalene	14.5 ± 2.0b	6.9 ± 1.7ab	37.9 ± 11.1a	13.0 ± 1.1c	9.1 ± 1. 8a	30.0 ± 10.5ab
2-Methylnaphthalene	11.5 ± 1.1b	1.5 ± 1.1c	8.0 ± 2.7b	13.5 ± 3.3c	6.5 ± 0.8abc	11.5 ± 8.4abc
2-Isopropylnaphthalene (beta-)	18.0 ± 2.4b	2.5 ± 1.1bc	1.2 ± 0.6c	22.5 ± 7.4bc	2.5 ± 0.3def	18.0 ± 9.2abc
2, 6-Diisopropylnaphthalene	24.5 ± 12.8b	0.1 ± 0.1c	4.5 ± 1.7bc	35.0 ± 7.0bc	2.8 ± 0.4def	0.0± 0.0c
2, 7-Diisopropylnaphthalene	19.5 ± 5.8b	1.8 ± 0.9bc	0.9 ± 0.4c	34.5 ± 8.3bc	3.8 ± 0.4cde	7.0 ± 4.7abc
F; MSD; df; P	95.42; 20.653; 11,108; < 0.0001	5.96; 5.406; 11, 108; < 0.0001	10.01; 6.117; 11, 108; < 0.0001	36.19; 23.770; 11,108; < 0.0001	22.59; 3.307; 11, 108; < 0.0001	3.94; 27.079; 11, 108; < 0.0001

[a]Data expressed as mean ± SE.

Acknowledgements

Srinidhireddy Kambham is thanked for his excellent technical assistance. We are indebted to Drs. James Ottea, William Crowe, Claudia Husseneder and Jianzhong Sun for reviewing this manuscript. This project was partially supported by USDA/ARS-New Orleans (58-6435-8-084) and approved for publication by the director of the Louisiana Agricultural Experiment Station as manuscript # 03-26-1441.

REFERNCES

1. Spink, W.T. The Formosan subterranean termite in Louisiana. Circ. No. 89, Louisiana Agricultural Experiment Station, Baton Rouge, LA 1967, pp. 12.
2. Mauldin, J.K., Jones, S.C., Beal, R.H. Soil termiticides: a review of efficacy data from field tests. *Internet Research Group on Wood Preservation*, 1987, No. IRG/WP/3666.
3. Su, N-Y and Scheffrahn, R.H. Formosan subterranean termite. 2000, Publication number: EENY-121, http://creatures.ifas.ufl.edu/urban/termites/formosan termite.htm, p. 5.
4. Henderson, G. Practical considerations of the Formosan subterranean termite in Louisiana: A 50-year-old problem. *Sociobiology*, **2001,** *37*, 281-292.
5. Katsura, E., Ogawa, H., Kojima, H., and Fukushima, A. Indoor air pollution by chlorpyrifos and S-421 after application for termite control. *Japanese Journal of Toxicology and Environmental Health*, **1996,** *42,* 354-359.
6. Carr, R.L., Ho, L.L., and Chambers, J.E. Selective toxicity of chlorpyrifos to several species of fish during an environmental exposure: biochemical mechanisms. *Environmental Toxicology and Chemistry*, **1997,** *16,* 2369-2374.
7. Kumar, A. and Chapman, J.C. Profenofos residues in wild fish from cotton-growing areas of New South Wales, Australia. *Journal of Environmental Quality*, **2001,** *30,* 740-750.
8. Sim, M., Forbes, A., McNeil, J., and Roberts, G. Termite control and other determinants of high body burdens of cyclodiene insecticides. *Archives of Environmental Health*, **1998,** *53,* 114-121.
9. Ibrahim, S. A., Henderson, G, Fei, H., and Laine, R.A. Toxic and repellent effects of 2'-Acetonaphthone on *Coptotermes formosanus* (Isoptera: Rhinotermitidae). *Sociobiology*, **2004,** *43,* 429-443.
10. Ibrahim, S. A., Henderson, G., Fei, H., and Laine, R.A. Survivorship, tunneling and feeding behaviors of *Coptotermes formosanus* (Isoptera: Rhinotermitidae) in response to 2'-Acetonaphthone-treated sand. *Pest Manag. Sci.* **2004,** *60,* 746-754.
11. Grace, J.K. and Yates, III J.R. Behavioral effects of a neem insecticide on *Coptotermes formosanus* (Isoptera: Rhinotermitidae) *Tropical Pest Management*, **1992,** *38,* 176-180.

12. Lin, T-S. and Yin, H-W. The effects of *Cinnamomum* spp. oils on the control of the termite *Coptotermes formosanus* Shiraki. *Taiwan Forestry Research Institute, New Series*, **1995,** *10,* 459-464.
13. Zhu, B.C.R., Henderson, G., Chen, F., Maistrello, L., and Laine, R.A. Nootkatone is a repellent for Formosan subterranean termite (*Coptotermes formosanus*). *Journal of Chemical Ecology*, **2001,** *27,* 523-531.
14. Berger, D. What ingredients are in mothballs? 1998. http://www.madsci.org/posts/archieves/may98/894550073.Gb.r.html.
15. Agency for Toxic Substances and Disease Registry (ATSDR) Toxicological profile for naphthalene, 1-methylnaphthalene and 2-methylnaphthalene. 1995. http://www.atsdr.cdc.gov/tfacts67.html, http://www.atsdr.cdc.gov/toxprofiles/phs67.html.
16. National Toxicological Program (NTP). Technical report on the toxicology and carcinogenesis studies of naphthalene. NTP TR 500, NIH Publication No. **01-4434**, 2000, pp. 11.
17. US Department of Health and Human Services (US-DHHS). Registry of Toxic Effects of Chemical Substances (RTECS, online database). 1993. National Toxicology Information Program, National Library of Medicine, Bethesda, MD.
18. Tingle, M.D., Pirmohamed, M., Templeton, E., Wilson, A.S., Madden, S., Kitteringham, N.R., and Park, B.K. An investigation of the formation of cytotoxic, genotoxic, protein-reactive and stable metabolites from naphthalene by human liver microsomes. *Biochemical Pharmacology*, **1993,** *46,* 1529-1538.
19. Wilson, A.S., Tingle, M.D., and Kelly, M.D. Evaluation of the generation of genotoxic and cytotoxic metabolites of benzo[a]pyrene, aflatoxin B1, naphthalene and tamoxifen using human liver microsomes and human lymphocytes. *Human and Experimental Toxicology*, **1995,** *14,* 507-515.
20. US Environmental Protection Agency (US-EPA). Integrated risk information system (IRIS). Toxicological review for naphthalene (CASRN 91-20-3). National Center for Environmental Assessment, Office of Research and Development, Washington, DC. 1998. http://www.epa.gov/iris/subst/0436.htm, pp. 24.
21. Fanucchi, M.V., Hinds, D., Cuellar, J.M., and Plopper, C.G. Species-specific susceptibility of neonatal mice and rats to naphthalene and 1-nitronapthalene. *American Journal of Respiratory and Critical Care Medicine*, **2000,** *161,* A172.
22. Chen, J., Henderson, G., Grimm, C.C., Lloyd, S.W. and Laine, R.A. Naphthalene in Formosan subterranean termite carton nests. *Journal of Agricultural Food and Chemistry*, **1998a,** *46,* 2337-2339.
23. Chen, J., Henderson, G., Grimm, C.C., Lloyd, S.W. and Laine, R.A. Termites fumigate their nests with naphthalene. *Nature*, **1998b,** *392,* 558-559.
24. Stowell, J.C. Composition and methods for killing termites. 1997. United States Patent # 5, 637, 298.
25. Green, III F., Kuster, T.A., Ferge, L. and Highley, T.L. Protection of southern pine from fungal decay and termite damage with N,N-

naphthaloylhydroxylamine. *International Biodeterioration and Biodegradation*, **1997,** *39,* 103-111.

26. Irwin, R.J., Mouwerik, M.V., Stevens, L., Seese, M.D., and Basham, W. Environmental contaminants encyclopedia, naphthalene entry. 1997. http://www.nature.nps.gov/toxic/naphthal.pdf. pp. 80.
27. Lovegreen, N.V., Fisher, G.S., Legendre, M.G., and Schuller, W.H. Volatile constituents of dried legumes. *Journal of Agricultural and Food Chemistry*, **1979,** *27,* 851-853.
28. Thompson, A.C., Hedin, P.A., Gueldner, R.C., and Davis, F.M. Corn bud essential oil. *Phytochemistry*, **1974,** *13,* 2029-2032.
29. Environmental Protection Agency (EPA) 2, 6-Diisopropylnaphthalene temporary exemption from tolerance requirement 9/99. 1999. http://pmep.cce.cornell.edu/profiles/herb-growthreg/24-d-butylate/26-Diisopropylnaphthalene/Diisopropylnap_tol_999.html. Vol. *64*, No. 183. September 22, 1999.
30. James, D.K., Komin, A.P., and Siegman, J.R. Process for preparation of 2-(6-methoxy-2-naphthyl) propionic acid and intermediates therefore utilizing 2, 6-diisopropylnaphthalene. 1994. US Patent No. 5,286,902.

31 Ministry of Agriculture, Forestry, and Fisheries (MAFF). Diisopropylnaphthalene in food packaging made from recycled paper and board, Food surveillance information sheet No.169. 1999.

32. Bloomquist, J.R. Insecticides: Chemistries and characteristics. 1996. http://ipmworld.umn.edu/chapters/bloomq.htm. Radcliffe's IPM World Textbook http://ipmworld.umn.edu, University of Minnesota, St Paul, MN. p .18.
33. Honda, T., Motoyama, M., Kiyozumi, M., and Kojima, S. Pulmonary toxicity of 2-isopropylnaphthalene and its photoproducts. *Eisei Kagaku*, **1991,** *37,* 300-306.
34. Huntingdon Life Science (1999) DIPN mammalian cell mutation assay. *Report RTG 001/984972,* February 3, 1999 and *RTG 002/994287,* November 12, 1999. http://archive.food.gov.uk/maff/archive/food/infsheet/1999/no169/169dipn.htm.
35. Abbott, W.S. A method of computing the effectiveness of an insecticide. *Journal of Economic Entomology*, **1925,** *18,* 265-267.
36. Finney, D.J. Probit analysis. Second edition. Cambridge University Press, Cambridge, MA. 1971. p. 256.
37. Delgarde, S. and Rouland-Lefevre, C. Evaluation of the effects of thiamethoxam on three species of African termite (Isoptera: Termitidae) crop pests. *Journal of Economic Entomology*, **2002,** *95,* 531-536.

38 SAS Institute. SAS/STAT User's Guide: Version 8, 1999. SAS Institute Inc., Cary, NC.

39. Karthikeyan, R. and Bhandari, A. Anaerobic biotransformation of aromatic and polycyclic aromatic hydrocarbons in soil microcosms, *Journal of Hazardous Substance Research*, **2001,** *3,* 1-19.
40. Knightes, C.D. and Peters, C.A. Substrate interactions in the biodegradation kinetics of polycyclic aromatic hydrocarbons (PAH) mixtures *Biotechnology and Bioengineering*, **2000,** *69,* 160-170.
41. Nussbaumer, P., Dorfstätter, G., Leitner, I., Mraz, K., Vyplel, H., and Stütz, A. Synthesis and structure-activity relationships of naphthalene-substituted derivatives of the allylamine antimycotic terbinafine. *Journal of Medicinal Chemistry*, **1993,** *36,* 2810-2816.
42. Abernathy, C.O., Hodgson, E., and Guthrie, F.E. Structure-activity relations on the induction of hepatic microsomal enzymes in the mouse by 1, 1, 1-trichloro-2, 2-bis (p-chlorophenyl) ethane (DDT) analogs. *Biochemical Pharmacology*, **1971,** *20,* 2385-2393.
43. Coats, J.R. Structure-activity relationships among DDT derivatives. *Journal of Environmental Science and Health*, **1983,** *18,* 173-188.
44. Ford, M.G., Greenwood, R., Turner, C.H., Hudson, B., and Livingstone, D.J. The structure/activity relationships of pyrethroid insecticides, 1. A novel approach based upon the use of multivariate QSAR and computational chemistry. *Pesticide Science*, **1989,** *27*, 305-326.
45. Konno, Y. and Shishido, T. A relationship between the chemical structure of organophosphates and insensitivity of acetylcholinesterase in the diamondback moth, *Plutella xylostella* L. *Applied Entomology and Zoology*, **1994,** *29,* 595-597.
46. Singh, A.K. Development of quantitative structure-activity relationship (QSAR) models for predicting risk of exposure from carcinogens in animals. *Cancer Investigation*, **2001,** *19,* 611-620.
47. Hudson, B.D., George, A.R., Ford, M.G., and Livingstone, D.J. Structure-activity relationships of pyrethroid insecticides, Part 2, The use of molecular dynamics for conformation searching and average parameter calculation. *Journal of Computer-Aided Molecular Design*, **1992,** *6*, 191-201.

48. Metcalf, R.L., Reinbold, K.A., Sanborn, J.R., Childers, W.F., Bruce, W.N., and Coats, J. Comparative biochemistry, biodegradability, and toxicity of DDT and carbofuran analogues. *WRC Research Report*, **1974,** *95*, 45.
49. Narahashi, T. Neuroreceptors and ion channels as the basis for drug action: past, present, and future. *Journal of Pharmacology and Experimental Therapeutics*, **2000,** *294*, 1-26.
50. Browning, H.C., Frase, F.C., Shapiro, S.K., Glickman, I., and Dubrule, M. Biological activity of DDT and related compounds. *Canadian Journal of Veterinary Research*, **1948,** *26D,* 282-300.

Chapter 4

Inhibition of *Blattella germanica* Acetylcholinesterase by *Bis*(n)-Tacrines: Prospects for the Molecular Design of a Selective Insecticide for a Household Pest.

[1]James M. Mutunga, [1]Troy D. Anderson, [2]Dawn M. Wong , [2]Paul C. Carlier, and [1]Jeffrey R. Bloomquist

Departments of [1]Entomology and [2]Chemistry, Virginia Polytechnic Institute and State University, Blacksburg, Virginia, U.S.A. 24061

The German cockroach (*Blattella germanica*, L.) is a major household pest that has developed resistance to most insecticides on the market. There is a need to develop insecticides that are less likely to induce resistance, are potent against insect pests including resistant populations, and possess less toxicity to humans. In this study, 9-amino-1,2,3,4-tetrahydroacridine (tacrine) was chosen as the acetylcholinesterase (AChE) inhibitor pharmacophore. We screened bivalent *bis*(n)-tacrines having methylene linkers from 2-12 carbons in length and determined their activity on *Bg*AChE, to probe the geometry of the AChE active site gorge. The dimeric tacrine having an octylene linker [*bis*(8)-tacrine] was the most potent analog against *Bg*AChE (IC_{50}= 68 nM). Some binding interference was observed with a 2-methylene linker ("C_2 bump") and with a 12-methylene linker ("C_{12} bump"), associated with a 4-fold and 7-fold loss in potency, respectively. It is possible that such "bumps" might convey underlying structural preferences of AChE in general, or of a particular species of AChE; further screening is on-going to test this hypothesis. Moreover, the most significant finding is that tether length dependent inhibition potency of *bis*(n)-tacrines relative to tacrine seems to differ across organisms, with the *Bg*AChE being both less sensitive overall, and less dependent on compound length compared to rat AChE. Such differences provide opportunities for comparative molecular modeling, and may inspire the synthesis of compounds that are specific and selective to insects.

Introduction

The German cockroach, *Blattella germanica* (Linnaeus), is a ubiquitous indoor pest that poses a direct human health risk throughout the United States. Cockroaches thrive in human households where excessive moisture, cracks and crevices, and abundant food sources are present (*1*). The U.S. Department of Housing and Urban Development reports asthma to be a prevalent residential health hazard to which cockroach allergens are the principal risk factors in asthma morbidity and mortality (*2-4*). Moreover, cockroaches are a potential vector of medically-important microorganisms including pathogenic bacteria or fungi, and parasites (*5-9*). Evidence of disease transmission is circumstantial in the US, but the threat is great enough to warrant control measures which will also reduce their nuisance and concomitantly their negative impact on the quality of human life.

There are several methods of German cockroach control using chemical insecticides, and insecticide resistance is common in this species. Metabolic resistance (elevated esterase, monooxygenase, and glutathione *S*-transferase enzymes) and target site insensitivity [e.g., altered sodium channel (*kdr*-type) and altered acetylcholinesterase ($AChE^R$)] have been documented (*10-19*). Continued successful control is at risk due to established resistance and therefore there is a need to develop new insecticides. The design of new AChE inhibitors is one possible way of sustaining chemical control as a principal component of IPM programs.

Acetylcholinesterase (EC 3.1.1.7) is found in synapses of most vertebrate and invertebrate species, where it terminates synaptic transmission by catalyzing the hydrolysis of the neurotransmitter, acetylcholine (ACh), to acetic acid and choline (*20*). Some insects have two AChE genes; *ace-1* and *ace-2* (e.g., mosquitoes and cockroaches) with only the *ace-1* gene being involved in nerve transmission (*21-26*) and therefore functionally important in insecticide action. Other insects like *Drosophila melanogaster* have only the *ace-2* gene (*27, 28*). *B. germanica ace* genes; *Bgace-1* and *B*gace-2, are orthologous to the insect *ace-1* and *ace-2* genes, respectively, and code for AChE-1 and AChE-2, respectively (*21*). In spite of the co-existence of the two genes in many insect species, the AChE-1 is the only type that has been associated with insensitivity to organophosphorus and carbamate insecticides and therefore plays a critical role as an insecticide target (*29*). Lack of potent chemical insecticides and resistance development pose a great threat of failure in pest control and there is a need to develop compounds that are more effective toward insect pests and less toxic to humans.

Carbamate and organophosphate insecticides inhibit the enzyme and cause excessive neuroexcitation due to accumulation of ACh at nerve synapses (*30*). The *Torpedo californica* AChE (*Tc*AChE), which is analogous to insect AChE-1, has two binding sites (Figure 1), the peripheral site at the mouth of a 20Å deep gorge and the catalytic site near the bottom (*31*).

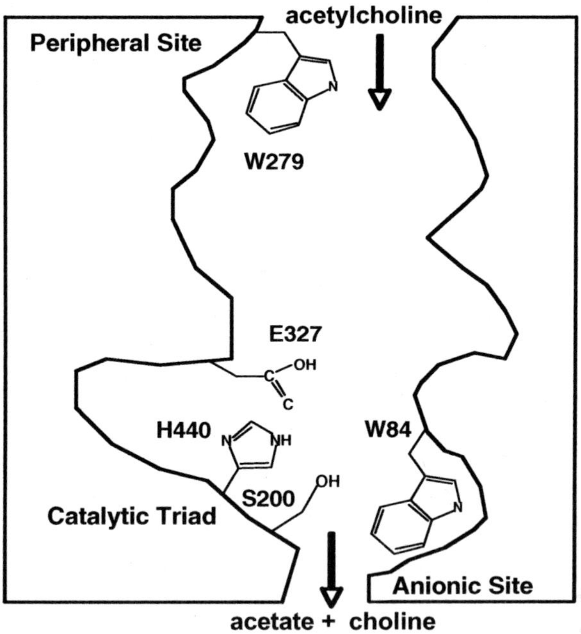

Figure 1. A simplified representation of the AChE gorge showing the peripheral aryl site and the catalytic active site with some key amino acids, numbered as in TcAChE. The substrate, acetylcholine, enters at the top of the gorge at the peripheral site (arrow), and the hydrolysis products exit near the bottom.

The *Tc*AChE gorge is lined with 14 conserved aromatic amino acids, which account for approximately 70% of the gorge residues (*32*). Aromaticity in the peripheral site is conserved across species, but may contain some unique residues, depending on the species (*33*).

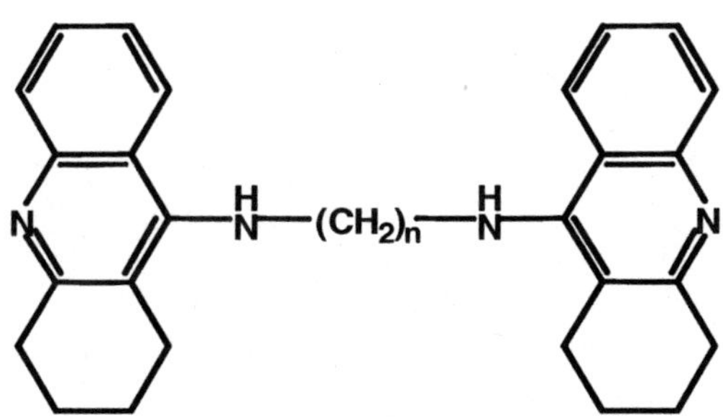

Figure 2. Chemical structure of the bis(n)-tacrines used in this study, with 'n' representing the number of carbons that form the tether linkage. The bis(n)-tacrines were synthesized and purified to >99.5% using established methods (34, 35).

Tacrine (9-amino-1,2,3,4-tetrahydroacridine) and its derivatives are potent inhibitors of human AChE (*h*AChE) and have been evaluated for their possible use as therapeutic agents to treat the memory loss of Alzheimer's Disease (*36, 37*). To achieve improved potency and selectivity through simultaneous binding to the active and peripheral site regions of AChE, methylene-linked tacrine dimers were prepared (*38*). The heptylene-bridged dimer *bis*(7)-tacrine indeed showed greatly enhanced potency and selectivity for inhibition of mammalian AChE compared to tacrine itself, which sparked considerable interest in the development of dimeric AChE inhibitors (*34, 39*). Subsequent X-ray crystallographic studies of *bis*(7)-tacrine and its short tether homolog *bis*(5)-tacrine complexed to *Torpedo californica* AChE (*Tc*AChE) experimentally verified bivalent binding of these inhibitors to the enzyme (*40*). In the present study, we investigated the inhibitory potency of the *bis*(n)-tacrine series (n = 2-10, 12) at *B. germanica*, to probe the differential inhibitory preferences of this enzyme and published data for rat AChE. The analysis of ligand-enzyme interaction at the active site informs the molecular design of potent bivalent insecticides against the cockroach and perhaps other insect pests and disease vectors.

Methods

Enzyme Activity Measurements

B. germanica adult females were obtained from the insecticide-susceptible CSMA strain of a laboratory colony maintained at the Department of Entomology, Virginia Polytechnic Institute and State University, Blacksburg, VA and stored at -80 ^{0}C before use. For enzyme inhibition assays, *B. germanica* heads were homogenized in 1ml of 0.1 M sodium phosphate buffer, pH 8.0, containing 1.5% Triton X-100 using a glass tissue homogenizer. Homogenates were centrifuged for 5 min at 10,000 × *g* and 4 °C in a Sorvall Fresco refrigerated centrifuge (Thermo Electronics Co., Germany), and the supernatant was used as the source of AChE for a 96-well microplate assay. AChE activity was measured by the method of Ellman (*41*), where the hydrolysis of acetylthiocholine (ACTh) was determined spectrophotometrically by absorbance of 5,5'-dithiobis-2-nitrobenzoic acid (DTNB)-thiocholine complex at 405 nm. The sodium phosphate buffer pH and Triton X-100 concentration requirements for optimal *Bg*AChE activity were determined empirically, and optimized ACTh and DTNB concentrations were 0.4 mM and 0.3 mM, respectively. All measurements were made at 25 °C in a DYNEX *Triad* microplate reader (DYNEX Technologies, Chantilly, VA, USA). We determined the potency of *bis*(n)-tacrines on *Bg*AChE by incubating the enzyme under different concentrations of the inhibitor (10^{-5} M-10^{-9} M) for 10 min in a total reaction volume of 200 μl per well. After addition of substrate and indicator, enzyme activity was monitored continuously at 405 nm for 10 min at 25 °C. Experimental sets without the inhibitor nor the enzyme and another containing the enzyme and substrate but without the inhibitor, were also included in each experiment as blanks and positive controls, respectively. Residual enzyme activities were converted to per cent of control and analyzed by nonlinear regression to a four parameter logistic equation to determine IC_{50} values and 95% confidence limits (CL) using Prism™ (GraphPad Software, San Diego, CA, USA).

Sequence Alignments

The catalytic subunits of *B. germanica* (*Bg*), *T. californica* (*Tc*), and *Rattus norvegicus* (*Rn*) full-length AChE sequences were aligned using CLUSTAL W (1.83) for multiple sequence alignments (*42*). Each alignment was cross-checked against crystallized, mature *Tc*AChE (PDB ID 2ace), as previously published (*43*). By convention, the main catalytic subunit of *Bg*AChE (Q1-N544) was numbered based on the mature form of *Tc*AChE, beginning at residue Q109 from the full-length sequence, Q2PZG3_BLAGE (*Bgace-1*).

Inhibition Profiles of AChE by *Bis*(n)-tacrines in Cockroach and Rat

Sigmoidal curves were obtained for each inhibitor (*e.g.*, Figure 3) and values of IC_{50} with 95% confidence limits were calculated by Prism software. Complete inhibition was achieved by both Propoxur and the non-covalent *bis*(n)-tacrine, which suggests that bona fide AChE-dependent hydrolysis of the acetylthiocholine substrate was the sole contributor to the enzyme activity measured in the head homogenates.

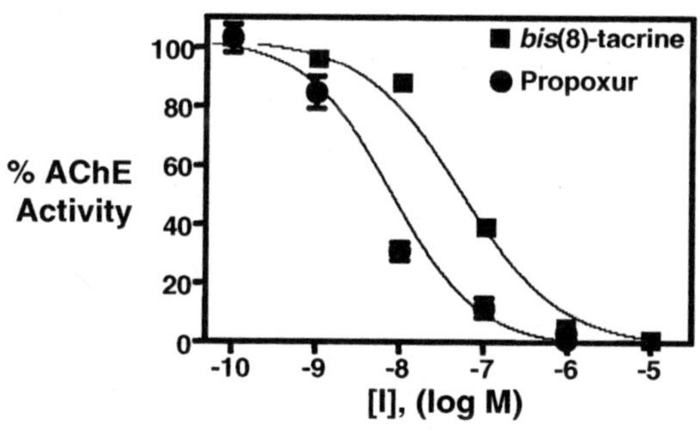

Figure 3. Representative sigmoidal plots showing the concentration-dependence of enzyme inhibition by bis(8)-tacrine and Propoxur against BgAChE.

The inhibitory effects of a range of tacrines tested on *Bg*AChE are summarized in Table 1. An increase in potency was observed with increased tether length up to *bis*(8)-tacrine, which had the highest potency. However, inhibitory potency of *bis*(8)-tacrine was nearly 14-fold less (Figure 3) than that of the commercial insecticide, Propoxur (IC_{50} = 5 nM). Potency then decreased as the tether length went from C_8 to C_{12} (Table 1). The dimer *bis*(8)-tacrine was

6- and 8-fold more potent compared to the monomer (tacrine) and *bis*(12)-tacrine, respectively. In contrast to the gain in potency between the monomer and *bis*(8)-tacrine, the sequential addition of methylene units in *bis*(9)- and *bis*(10)-tacrine reduced potency in approximately 2-fold steps (Table 1). Ultimately, there was a 19-fold loss in potency at the longest tether length ($n = 12$) compared to the optimum ($n = 8$). As expected, tether length can be suboptimal by being either too short or too long. Overall, there were "bumps" at C_2 and C_{12} where the effectiveness of the inhibitors declined. These "bumps" are best appreciated visually by viewing the bar graphs of Figure 4.

Table 1. Enzyme-ligand inhibition parameters for bivalent tacrines on *Bg*AChE and *Rn*AChE generated by four-parameter nonlinear regression analysis.

Compound	[a]Rn*AChE* IC_{50}*, nM* (± *SE*)	Bg*AChE* IC_{50}*, nM* *(95% CI)*	[b]*Selectivity: Rat/Cockroach*
Tacrine	223 (11)	394 (345-449)	0.6
[c]*bis*(2)-tacrine	711 (25)	1,699 (1,308-2,207)	0.4
bis(3)-tacrine	254 (55)	835 (768-908)	0.3
bis(4)-tacrine	157 (23)	529 (490-571)	0.3
bis(5)-tacrine	28 (5)	326 (257-415)	0.08
bis(6)-tacrine	3.8 (0.4)	196 (183-210)	0.02
bis(7)-tacrine	1.5 (0.3)	138 (112-170)	0.01
bis(8)-tacrine	7.8 (0.9)	68 (67-73)	0.11
bis(9)-tacrine	31 (3)	131 (110-157)	0.24
bis(10)-tacrine	40 (6)	200 (161-249)	0.2
bis(12)-tacrine	[d]N/A	1,324 (1,267-1,383)	---

[a]Data taken from Carlier *et al* 1999 (*44*), using AChE from rat cortex homogenate.

[b]Selectivity is defined as rat IC_{50}/cockroach IC_{50}.[b]

[c]Numbers *n* in the term *bis*(n)-tacrine represents the number of methylene units in the linker that tethers the tacrine moieties.

[d]N/A means no data are available for *bis*(12)-tacrine in *Rn*.

Inspection of Figure 4 and Table 1 reveals both similarities and differences in the responses of cockroach and rat enzymes to inhibition by *bis*(n)-tacrines. Both enzymes are most potently inhibited at a tether length of 7-8 carbons. Both have a "bump" at a C_2 tether length, and while no data are available for *bis*(12)-tacrine in the rat, inhibitory potency declines in both species as the tether length exceeds the optimal carbon chain length. There is also a more steep tether length dependence in the rat compared to cockroach, since potency declines about 3-fold with a 2 carbon change from optimal tether length in *Bg*AChE, while potency declines about 20-fold with a 2 carbon change from optimal tether length in *Rn*AChE. This difference suggests more efficient dual site binding in the rat enzyme. Finally, the *bis*(n)-tacrines are uniformly more active against

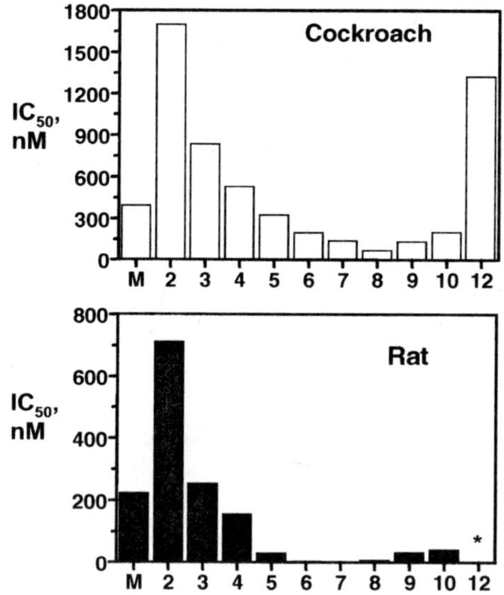

Figure 4. Plots of tether length dependent potency of bis*(n)-tacrines on insect,* Blattella germanica *(A) and vertebrate,* Rattus norvegicus *(B) AChE. Data for rat AChE are from reference (44). *Indicates no data available for* bis*(12)-tacrine. Note the difference in Y-axis scale. M = tacrine monomer, and numbers indicate tether length in carbon atoms (e.g., 2 =* bis*(2)-tacrine, etc.).*

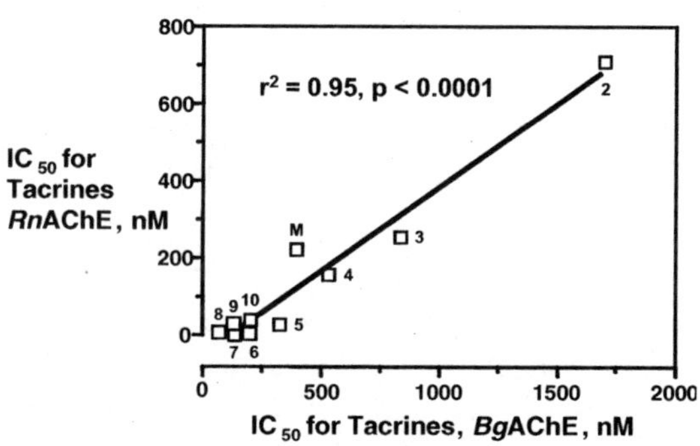

*Figure 5. Correlation plot of tether length dependent potency of dimeric tacrines on insect, (*Bg*AChE) and vertebrate (*Rn*AChE) AChE. Data extracted from Table 1. M = tacrine monomer, and numbers indicate tether length in carbon atoms (e.g., 2 =* bis*(2)-tacrine, etc.).*

*Rn*AChE than *Bg*AChE, at all tether lengths, as reflected in calculated selectivities that are <1 (Table 1), but there is a highly significant overall correlation in responses between the two species (Figure 5).

Protein Structural Interpretation of Results

Protein sequence alignments (Figure 6) show a high degree of homology between cockroach and vertebrate AChEs at identified structural motifs, and suggest a number of explanations for the observed data. The "bump" at n=2 in both species is very likely due to steric crowding at the bottom of the gorge, where the close attachment of the second bulky tacrine unit disrupts the preferred binding mode of the catalytic site tacrine. A similar argument was made by Carlier et al. in the original study of *bis*(2)-tacrine inhibition of *Rn*AChE (*34*). In complexes of *bis*(5)- and *bis*(7)-tacrine with *Tc*AChE, one tacrine unit binds to the choline-binding site (Figure 6), sandwiched between the aromatic side chains of W84 and F330, at the bottom of the active-site gorge, as seen in the structures of monomeric tacrine (PDB ID 1ACJ) (*44*) and other tacrine-based bivalent inhibitors in complex with *Tc*AChE (PDB ID 1ODC, 1UT6, 2CEK) (*45, 46*). We conclude that a similar mechanism operates in *Bg*AChE.

The X-ray crystal structure of *bis*(7)-tacrine complexed to *Tc*AChE also reveals a π-complex sandwich of the peripheral site tacrine unit with W279 and Y70; comparison of the *Tc*AChE and *Rn*AChE sequences suggests the identical binding mode would be realized for *Rn*AChE. Lower overall potency of tacrine dimers against *Bg*AChE relative to *Rn*AChE may be due, in part, to the presence of I73 instead of the Y70 found in the vertebrate peripheral site (Figure 6). Such a substitution would diminish π-π/cation-π interaction at the peripheral site of *Bgace-1* with the second tacrine moiety. Despite the loss of binding interaction, based on the I73 (*Bgace-1*) replacement, the other peripheral site residue Y336 (*B*gace-*1*), corresponding to G342 and G335 (*Tc*) may give additional complementary protein-ligand interaction for the longer tether length *bis*-tacrines (A6A to A10A). The C_2 and C_{12} "bumps" were associated losses in potency, suggesting that steric interference occurs in binding midway and at the mouth of the gorge. The interfering residues or additional mechanisms are yet to be identified.

A recent study showed the possibility of steric interference mediating the binding of dimeric tacrines in *Torpedo californica* (*46*). In that study, *bis*(5)-tacrine was shown to bind to W84 and F330 at the catalytic site of *Tc*AChE, but because of the shorter tether length, it does not interact with W279 at the peripheral site. Instead, the *bis*(5)-tacrine displaces the phenyl residue of F331 (Figure 6), causing a major rearrangement in the W279-S291 loop, and thereby inducing a major rearrangement in the enzyme active site. A previous report has shown that C289 and R339 are unique to certain insects (including *B. germanica*) and are found within the gorge (*33*). In particular, the possibility of alkylating C289 was hypothesized (*33*), but other investigators have concluded that the thiol may be shielded by adjacent amino acids that prevent this reaction (*47*).

	catalytic triad			*oxyanion hole*		*choline-binding site*	
	200	327	440	116	201	84	330
Tc	TIFGESAGGAS	E	H	YGGGF	A	W	FF
	203	334	447	119	204	86	337
rat	TLFGE**S**AGAAS	E	H	YGGGF	A	W	YF
	202	328	442	119	203	87	331
Bg	TLFGE**S**AGAVS	E	H	FGGGF	A	W	YF

	acyl pocket			*peripheral site*				
	233	288	290	70	72	121	279	334
Tc	W	F	F	Y	D	Y	W	YG
	236	295	297	72	74	124	286	341
rat	W	F	F	Y	D	Y	W	YG
	235	289	291	73	75	124	283	335
Bg	W	**C**	F	**I**	D	Y	W	Y**Y**

*Figure 6. Alignment of T. californica (Tc), R. norvegicus (rat), and B. germanica (Bg) AChE. SwissProt codes are: ACES_TORCA (*Tc*); ACES_RAT (rat); Q2PZG3_BLAGE (*Bgace-1*). Residues marked in* ***bold*** *are the conserved catalytic serine, as well as other residues in the acyl pocket and peripheral site that differ in* B*gace-1 (see text for explanation). By convention, numbering is based on that of the catalytic subunit/mature form of* Tc*AChE, as defined by X-ray structures of the protein (*e.g., *PDB ID 2ace).*

Conclusions

The most potent *bis*(n)-tacrine in the cockroach, *bis*(8)-tacrine, was over 10-fold less potent than the commercial Blatticide, Propoxur. However, we do not view the *bis*(n)-tacrines as lead compounds, since they have no contact activity against insects, owing to the presence of basic nitrogens that bestow unfavorable pharmacokinetics (data not shown). Our studies found a greater overall potency for these compounds in vertebrates than *Bg*AChE. The greater potency against vertebrate AChE compared to *Bg*AChE suggests some structural/functional differences between insect and vertebrate AChEs and creates possibilities for further structure-activity investigation. This speculation could be further investigated by using *Ala*-scanning site directed mutagenesis of residues thought to interact with tacrines at the C_2 bottleneck and the peripheral site, and analyzing the recombinants for gorge geometry. As discussed earlier, some amino acid residues at the AChE gorge differ across species, and act in concert with each other to modulate interactions with ligands. It is possible that these interactions differ across species and therefore would form targets for the design of selective insecticides.

References

1. Brenner, B. L; Markowitz, S; Rivera, M; Romero, H; Weeks, M; Sanchez, E; Deych, E; Garg, A; Godbold, J; Wolff, M. S; Landrigan, P. J; Berkowitz, G. *Environ. Health Perspect.* **2003**, *111*, 1649-1653.
2. Rosenstreich, D. L; Eggleston, P; Kattan, M; Baker, D; Slavin, R. G; Gergen, P. *N. Engl. J. Med.* **1997,** *336*, 1356-1363.
3. Wang, C; Abou El-Nour, M. M; Bennett, G. W. *J. Community Health* **2008**, *33*, 31-39.
4. Pollart, S. M; Smith, T. F; Morris, E. C; Gelber L. E; Platts-Mills, T. A; Chapman M. D. *J. Allergy Clin. Immunol.* **1991**, *87*, 505–510.
5. Gergen, P. J; Mortimer, K. M; Eggleston, P. A; Rosenstreich, D; Mitchell, H; Schal, C; Hamilton, R. L. *Annu. Rev. Entomol.* **1990**, *35*, 521-51.
6. Schal, C; Gautier, J. Y; Bell, W. J. *Biological Reviews* **1984**, *59*, 209–254.
7. Ash, N; Greenberg, B. *J Med Entomol.* **1980**, *17*, 417-423.
8. Kopic, R. J; Sheldon, B; Wright, C. G. *J. Food. Protect.* **1994**, *57*, 125-132.
9. Dong, K; Scott, J. G. □*Med. Vet. Entomol.* **1992**, *6*, 241–243.
10. Appel, A. G. *J. Econ. Entomol.* **2003**, *96*, 863-870.
11. Lee, C. Y; Hemingway, J; Yap, H; Chong, N. L. *Med. Vet. Entomol.* **2000**, *14*, 11-18.
12. Xua, Q; Wang, H; Zhang, L; Liu, N. *Gene* **2006**, *379*, 62-67.
13. Scott, J. G. In *Pesticide Resistance in Arthropods;* Roush, R.T., Tabashnik, B. E. (Eds.); Chapman and Hall, New York, 1990; pp 39–57.
14. Zhai, J; Robinson, W. *Jpn J. Sanit. Zool.* **1991**, *42*, 241-244.
15. Hemingway, J; Field, L; Vontas, J. *Science* **2002**, *298*, 96–97.
16. Lee, C. Y; Yap, H. H; Chong, N. L. *Bull. Entomol. Res.* **1996**, *86*, 675- 682.
17. Fournier, D; Bride, J. M; Hoffmann, F; Karch, F. *J. Biol. Chem.* **1992**, *267*, 14270–14274.
18. Cochran, D. G. In *Understanding and controlling the German cockroach.* Oxford University Press: New York, 1995; pp 171-192.
19. Collins, W. J. *Pesticide Science* **2006**, *6*, 83-95.
20. Toutant, J. P; Arpagaus, M; Fournier, D. *J. Neurochem.* **1988**, *50*, 209–218.
21. Kim, J. I; Jung, C. S; Koh, Y. H; Lee, S. H. *Insect Mol. Biol.* **2006**, *15*, 513-522.
22. Bourguet, D; Raymond, M; Fournier, D; Malcolm, C. A; Toutant, J. P; Arpagaus, M. *J. Neurochem.* **1996**, *67*, 2115–2123.
23. Weill, M; Lutfalla, G; Mogensen, K; Chandra, F; Berthomieu, A; Vertical, C; Pasteur, N; Philips, A; Fort, P; Raymond, M. *Nature* **2003**, *423*, 136–137.
24. Nabeshima, T; Kozaki, T; Tomita, T; Kono, Y. *Biochem. Bioph. Res. Co.* **2003**, *307*, 15–22.
25. Nabeshima, T; Mori, A; Kozaki, T; Iwata, Y; Hidoh, O; Harada, S; Kasai, S; Severson, D. W; Kono, Y; Tomita, T. *Biochem. Bioph. Res. Co.* **2004**, *313*, 794–801.
26. Grauso, M; Culetto, E; Combes, D; Fedon, Y; Toutant, J. P; Arpagaus M. *FEBS Lett.* **1998**, *424*, 279–284.
27. Baek, J. H; Kim, J. I; Lee, D. W; Chung, B. K; Miyata, T; Lee, S. H. *Pesticide Biochem. Physiol.* **2005**, *81*, 164–175.

28. Weill, M; Fort, P; Berthomieu, A; Dubois, M. P; Pasteur, N; Raymond, M. *Proc. R. Soc. Lond. Ser. B: Biol. Sci.* **2002**, *269*, 2007–2016.
29. Fournier, D; Mutéro, A. *Comp. Biochem. Physiol.* **1994**, *108*, 19–31.
30. Coppage, D. L; Matthews, E. *Bull. Environ. Contam. Toxicol.* **1974**, *11*, 483–488.
31. Sussman, J. L; Harel, M; Frolow, F; Oefner, C; Goldman, A; Toker, L; Silman, I. *Science* **1991**, *253*, 872-879.
32. Axelsen, P. H; Harel, M; Silman, I; Sussman, J. L. *Protein Sci.* **1994**, *3*, 188-197.
33. Pang, Y-P. PLoSONE. **2006**, *1*: e58. doi:10.1371/journal.pone.0000058.
34. Carlier, P. R; Han,Y. F; Chow, E. S-H; Li, C. P-L; Wang, H; Lieu, T. X; Wong, H. S; Pang, Y-P. *Bioorg. Med. Chem.* **1999**, *7*, 351-357.
35. Wang, H; Carlier, P; Ho, W; Wu, D; Lee, N; Li, C; Pang, Y; Han, Y. *NeuroReport* **1999**, *10*, 789-793.
36. Relman, A.S. *N. Eng. J. Med.* **1991**, *324*, 349.
37. Gregor, V.E; Emmerling, M.R; Lee, C; Moore, C.J. *Bioorg. Med. Chem. Lett.* **1992**, *2*, 861-864.
38. Pang, Y.-P; Quiram, P; Jelacic, T; Hong, F; Brimijoin, S. *J. Biol. Chem.* **1996**, *271*, 23646-23649.
39. Du, D. M; Carlier, P. R. *Curr. Pharm. Des.* **2004**, *10*, 3141-3156.
40. Rydberg, E.H; Brumshtein, B; Greenblatt, H.M; Wong, D.M; Shaya, D; Williams, L.D; Carlier, P.R; Pang, Y.P; Silman, I;Sussman, J.L. *J. Med. Chem.* **2006**, *49*, 5491-5500.
41. Ellman G. L; Courtney, K. D; Andres, V; Featherstone, R. M. *Biochem. Pharmacol.* **1961**, *7*, 88-95.
42. EMBL-EBI ClustalW2. http://www.ebi.ac.uk/Tools/clustalw2/index.html.
43. Raves, M. L; Harel, M; Pang, Y-P; Silman, I; Kozikowski, A. P; Sussman, J. L. *Nat. Struct. Biol.* **1997**, *4*, 57-63.
44. Harel, M; Schalk, I; Ehret-Sabatier, L; Bouet, F; Goeldner, M; Hirth, C; Axelsen, P. H; Silman, I; Sussman, J. L. *Proc. Natl. Acad. Sci. USA.* **1993**, *90*, 9031-9035.
45. Colletier, J. P; Sanson, B; Nachon, F; Gabellieri, E; Fattorusso, C; Campiani, G; Weik, M. *J. Am. Chem. Soc.* **2006**, *128*, 4526-4527.
46. Rydberg, E. H; Brumshtein, B; Greenblatt, H. M; Wong, D. M; Shaya, D; Williams, L. D; Carlier, P. R; Pang, Y-P; Silman, I; Sussman, J. L. *J. Med. Chem.* **2006**, *49*, 5491-5500.
47. Rowland, M; Tsigelny, I; Wolfe, M; Pezzementi, L. *Chemico-Biolog. Inter.* **2008**, Epub, citation in press.

Chapter 5

Screening insecticides for use as soil termiticides requires a series of bioassays: lessons from trials using *Reticulitermes flavipes* (Isoptera: Rhinotermitidae).

Incorporating termite behavior into termiticide bioassay design.

Brian T. Forschler

Department of Entomology, University of Georgia, Athens, GA 30602

Tests of soil termiticide efficacy should consider termite behavioral reaction to the toxicant. This chapter outlines a series of four separate bioassays that account for factors such as bioavailability, dose acquisition, route of entry, movement of intoxicated termites, and collective decision making that are important in modeling termite population response to field application. Results with seven different formulated termiticides indicate that the moniker repellent or non-repellent are a function of concentration not the purview of any particular class of chemistries. Bioavailability was evident in that mortality decreased with all active ingredients tested in bioassay with sandy loam soil compared to play sand with one exception, imidacloprid. Data included evidence that acquisition of a dermal dose is slow, requiring over one hour continuous exposure to treated sandy loam soil to produce >50% mortality at 10 ppm with all the formulations tested. Dermal and oral toxicities varied within all termiticides but one – fipronil - and all but one, the chlorfenapyr formulation, affected worker termite performance on an experimental trail compared to controls. Three behavioral conditions that must be considered when interpreting termiticide efficacy bioassay data are also discussed. First, the main route of entry for subterranean termites confronted with a soil-based insecticide is oral. Second, assignment of the task of gallery construction to specific workers and third, a collective decision making system of communication used by worker termites traveling through a network of galleries. It is therefore unlikely that consistent transfer of soil-borne toxicants will follow field application of any termiticide. A choice bioassay design is suggested as a more realistic approximation of termiticide field efficacy.

The process of protecting structures from subterranean termite infestation essentially involves breaking the chain of events and lines of communication that termites use to locate and establish feeding sites (*1*). Termite control tactics can be soil- or structure-based using methods such as termite resistant building materials, physical barriers, chemical barriers, or population management practices (*1, 2*). Application of a chemical barrier, however, has been the most common practice over the last 60 years (*2*). This is accomplished by placing an insecticide (termiticide) into the soil surrounding a structure for the purpose of excluding termites.

The screening of candidate insecticides for use with the exclusionary hypothesis employed in the last half of the 20th century was characterized by Ebeling and Pence (*3*) who determined efficacy in a constant exposure bioassay with the acceptable candidate producing rapid mortality. Twenty years later, Su et al. (*4*) demonstrated the need to incorporate termite behavioral reaction to the chemistry being tested in bioassay. This seminal work moved termiticide efficacy away from constant exposure mortality assays to experiments that allowed termites to construct galleries through treated soil in a "qualitative" tube bioassay system, which has become a commonly used evaluation tool (*5, 6, 7*). Su et al. (*4*) also defined the behavioral reaction of termites, using a petri dish (qualitative) assay, along a gradient as either 1 (I) – repellent, 2 (II) not repellent, or 3 (III) slow-acting stomach poison.

The termiticide active ingredients used between 1989-96 (several synthethic pyrethroids – cypermethrin, permethrin, bifenthrin, and two organo-phosphate isophenphos and chlorpyrifos) conformed to the exclusionary hypothesis. As defined by the tube bioassay, these termiticides, through rapid mortality or 'repellence', eliminated termite traffic into treated soil (*8, 9, 10, 11*). The so-called 'repellent termiticides' were envisioned to alter foraging, leaving the resident termite population unaffected and free to search for food resources. The lack of termite population impact implies that in the field, continued foraging will eventually, through application error, construction fault, or degradation of the active ingredient, allow termites to locate and exploit an untreated area around the structure/soil interface returning that structure to the list of active feeding sites – i.e. infestation.

In 1996, Bayer Environmental Science introduced a new class of chemistry, the neonicatinamides, that provoked a paradigm shift from the exclusion hypothesis to that of the 'treated zone'. That product, Premise®, with the active ingredient imidacloprid, was described as neither repellent nor fast acting at label application rates and allowed subterranean termites to excavate galleries into the treated soil, bringing the term 'non-repellent termiticide' into the lexicon. Since 1996, two additional 'non-repellent' products have gained Environmental Protection Agency (EPA) registration, Termidor® (fipronil) and Phantom® (chlorfenapyr). Theoretically, use of a non-repellent termiticide allows termites to construct galleries through the treated soil and provides structural protection through attrition within the offending termite population. The attrition hypothesis has gained acceptance in the termite control industry due to successful field trials reporting termite population impacts following treatment (*12, 7*).

In order to fulfill the promise of the attrition hypothesis, a non-repellent termiticide should display delayed mortality and horizontal transfer between nest mates following contact with a lethal dose (*13*). The hypothesis of non-repellent termiticide efficacy, under the construct of attrition, predicts that structural protection is accomplished through population reduction achieved by transfer of a lethal dose through direct contact with treated soil or behavioral interactions (contact, grooming, or food exchange). Attaining structural protection with 'non-repellent' chemistries is predicated on termites tunneling through the treated soil and acquiring an effective dose of the active ingredient. Therefore, the question of termiticide efficacy under the attrition hypothesis can be tested in bioassay by examining five critical elements: concentration of active ingredient, time of exposure, lethal dose, route of entry, and behavior of an affected termite.

Field efficacy of termiticides, under the attrition scenario, could be predicted in bioassay with the most efficacious active ingredient being one which kills termites at the lowest concentration following the shortest exposure while not affecting the behavior of the intoxicated subject. The behavioral component is required to facilitate transfer and/or maintain communication, and therefore continued movement, to the 'lethal zone'.

I decided to examine the questions of concentration, time of exposure, lethal dose, route of entry, and termite mobility to test the hypothesis that termiticide efficacy could be predicted from bioassay under the attrition hypothesis. This manuscript reports data from five separate bioassays. Six non-repellent termiticide formulations were tested at four concentrations and four exposure time periods in two soil types to obtain Lethal Concentration values by time of exposure, termite excavation through treated soils were tested in a choice test bioassay system. Lethal Dose values were generated examining oral and dermal routes of entry, and movement of dermal-dosed termites was timed. Results are discussed in regard to predicting field efficacy assuming structural protection through termite population reduction.

Materials and Methods

Termites

Eastern subterranean termites, *Reticulitermes flavipes* (Kollar), were collected from infested logs at the University of Georgia Whitehall Forest in Athens, GA using extraction methods as described by La Fage et al. (*14*) and modified by Forschler and Townsend (*15*). Termite colonies were identified to species using soldier characteristics (*16*). Termites collected from logs were placed in clear plastic boxes (26 × 19 × 9 cm) containing moistened 9-cm No. 1 Whatman filter papers and several thin pieces of pine (11.25 × 3.75 cm and 1 mm thick). Plastic boxes with termites were maintained in an environmental chamber at 24 °C for no more than one month prior to beginning a bioassay. Only undifferentiated *R. flavipes* workers, fourth instar and older, were used in bioassay.

Termiticides

The five termiticide formulations tested in these trials were Chlorfenapyr (Phantom TM, BASF, Parsippany, NJ), Thiamethoxam (CGA, 25 WG, Syngenta, Greensboro, NC), Imidacloprid (Premise 75, Bayer, Kansas City, MO), Fipronil (Termidor, 80WG, BASF, Research Triangle, NC) and indoxacarb (Steward, 15 SC, DuPont, Wilmington, DE).

Soil Exposure-Time Bioassay

Treatment of Soil/Sand

Termiticide concentrations were determined by calculating the amount of active ingredient needed to reach a concentration of 10,000 parts per million (ppm, w of AI/w of soil) when 20 ml of solution were added to 100g of soil. Subsequent concentrations were reached by serial dilution of the aforementioned 10,000 ppm solution. Termiticides were tested in one of two substrates; Cecil series sandy loam soil (71% sand, 21% silt, 8% clay) or play sand purchased commercially (100% sand).

Termiticide solutions, prepared as previously described to obtain the desired concentration, were added to 100g of substrate to reach 20% soil moisture and the appropriate solution slowly added to the substrate in a plastic bag (16.5 × 14-cm). The solution/substrate was thoroughly mixed by hand, through the bag, until all of the substrate was evenly moistened. Untreated control substrates were brought to 20% moisture using distilled water only. The moistened substrate, for each solution/time/concentration combination tested, was then evenly divided (≈ 33 g) among three 9-cm petri dishes and spread to form a continuous layer on the bottom using stainless steel spoons. Each petri dish was labeled with the termiticide solution used and concentration. Five concentrations (0.1, 1, 10, 100, and 1000 ppm) of each termiticide were tested and replicated at least 10 times per chemistry and substrate.

Exposure Time Period

Each termiticide and concentration was tested at four exposure time periods, 1, 10, 100, and 720 minutes. Petri dishes for the overnight exposure (720 minutes) treatments were placed in a plastic box (25 × 32.5 × 9-cm) and kept in an environmental chamber at 27 °C until time to remove the termites. Paper towels saturated with distilled water were placed in the bottom of the plastic containers to maintain high humidity conditions inside the overnight exposure arena. All other exposure times were maintained at room temperature.

One petri dish was used for all exposure times with a particular chemical and concentration combination for a single replicate. Ten termites were placed in the treatment petri dish at a particular termiticide concentration for the designated amount of time. There were 15 replicates performed for each solution/time/concentration combination. At the end of the time period termites were removed using featherweight forceps and placed into an observation petri dish (6 × 1.5-cm) that contained a 5.5-cm piece of No.1 Whatman filter paper moistened with 0.25-ml of distilled water. All observation petri dishes for each termiticide/concentration/time exposure combination were placed in separate plastic boxes (26 × 19 × 9-cm) in an environmental chamber at 27 °C. Paper towels saturated with distilled water were placed in the bottom of the plastic boxes to maintain high humidity. The number of living termites in each observation petri dish was counted every day for ten days to determine survivorship. All dead termites were removed to prevent transfer of toxicant due to cannibalism. Death was defined as lack of movement when touched by a probe.

Excavation Choice Bioassay

Nine round plastic containers (5-cm ID, 3.5-cm H) were connected using a 7-cm length of tygon tubing (2-mm ID). The central container was filled to a depth of 2.5-cm with a sand and vermiculite mixture (14:12 ratio) to provide a moisture-filled tunneling substrate and serve as the introduction chamber with the tubing entering the chamber at a height equivalent to the top of the sand/vermiculite mixture. The introduction chamber was connected to the base of four chambers, termed substrate chambers, containing play sand that was treated as described in the Time-Exposure section. Only one of the four substrate chambers, within any replicate, contained a treatment such that the termites introduced into each arena had a choice of three untreated and one treated substrate chamber. The four substrate chambers were connected by a 7-cm length of tygon tubing (placed on the opposite side from the tube leading into that chamber) to the base of another chamber, termed the food chamber, containing a single block (2-cm^3) of pine wood.

The arrangement of chambers resembled a wheel with the introduction chamber at the center with four spokes (tube-defined paths) each leading to a separate substrate chamber with access to a final food chamber. Each bioassay arrangement of nine chambers was considered one replicate. Five hundred termites were placed into the introduction chamber at the start of the bioassay and confined in that chamber for 24-h using small (1.9-cm width) binder clips (Charles Leonard Inc., Glendale, NY). The binder clips were positioned on the tubes leading from the introduction chamber near the point of attachment to the introduction chamber to provide a period of acclimatization prior to release into the choice arena. Termites from a single laboratory culture were used for each replicate. At least three different termite colonies (laboratory cultures) were used in the 6-16 replicates that composed this series of tests. Termiticides were tested at 50-60 ppm's to simulate approximate labeled application concentrations.

Route of Exposure Bioassay

Individual termites were held under a binocular dissecting microscope using a vacuum venturi system. The system consisted of a Pasteur pipette attached to the end of a 3-mm ID piece of tygon tubing which was attached using a plastic t-connection to an open length of tubing at one end and a vacuum source at the other. Termites were picked up by placing the Pasteur pipette tip on the abdomen while placing a finger over the open tube to create a vacuum at the pipette tip. Termites were then treated using one of several concentrations of the appropriate termiticide solution using a micro-applicator. Termites were released following treatment by removing the finger from the open tube to break the vacuum suction at the pipette tip.

Termites were treated and maintained following treatment using one of four scenarios. Termites were treated by placing 0.15 microliters on the pronotum, which was allowed to dry before they were placed in either a Petri dish containing 9 other similarly treated termites or isolated in an individual tissue culture well in a standard 96-well plate. Both the Petri dish and tissue culture wells were lined with an appropriate disk of untreated #1 Whatman filter paper moistened with de-ionized water. These treatment regimes simulated a dermal exposure while allowing for grooming by nest mates (the 10 termites in a Petri dish) and no grooming (isolated tissue culture well termites). The oral route of exposure was simulated by using the vacuum venturi system and microapplicator to place 0.125 microliters of the appropriate concentration on the mouthparts of each termite and only including those where the droplet disappeared into the bucal cavity (not splayed across the head capsule or mouthparts) and assumed to have been consumed. Termites treated in the oral route assays were held in Petri dishes with 9 other likewise treated individuals or as isolated individuals as described for the topical assays. Mortality was recorded daily for 10-17 days all dead termites were removed daily from Petri dishes or tissue culture wells.

Termite Running Assays

Termites follow ink lines drawn by pens containing 2-phenoxyethanol, such as Papermate® pens (*17*). Photocopies were made of a sheet of paper that had four straight, 6 cm-long lines with a 1 cm diameter circle drawn at one end of each line. Seconds prior to performing the assay one line was traced with a disposable Papermate® pen. One termite from a group that had been treated with a topical dose close to the LD_{50} was gently placed inside the 1-m circle using the vacuum venture system previously described. As soon as the termite started running along the straight ink line it was timed over the distance of 6-cm with a hand-held stopwatch. Speed was recorded only when termites ran 6 cm without stopping or straying from the line. After running, the termite was placed in one well of a standard 90-well tissue culture plate which contained a piece of #1 Watman filter paper and labeled as to the chemistry, dose, and day when it was treated. Termites were timed one hour after treatment and for four consecutive days thereafter. The tissue culture plate was placed inside a plastic box containing wet paper towels and maintained in an environmental chamber at 24 °C. A new pen line was drawn for each termite tested.

Data Analysis

No replicate in any bioassay was included in analysis if the control survivorship was equal or less than 90%. Probit analysis was used to calculate LD_{50} values from topical and force-feeding assays for examination of route of entry. Probit analysis also was used to calculate LC_{50} values for each combination of termiticide concentration and exposure period (*18*). Mean corrected percent mortality was compared by time and concentration within substrate type by termiticide using Log10 transformed data with the General Linear Models Procedure (*18*). Mean separation was accomplished using Protected Least Significant Difference (*18*).

No statistical comparisons were made between termiticide formulations because the active ingredients represented different modes of action and various concentrations were often tested to obtain the appropriate approximation of a lethal dose. Chlorfenapyr is a metabolic inhibitor that affects electron transport in the mitochondria. The remaining three insecticides are nerve toxins. Acetamiprid, Thiamethoxam and Imidacloprid affect acetyl choline receptors, Fipronil impacts GABA-gated chloride channels, Bifenthrin is a sodium channel modulator, and Indoxacarb blocks voltage dependent sodium channels. Therefore, analysis was confined to comparisons within termiticide formulations.

Results

Timed exposure bioassay

The only treatment that provided a consistent statistically significant response within soil type was thiamethoxam with decreased LC_{50} values as exposure time increased in both sand and sandy loam soil (Table I).

Chlorfenapyr provided LC_{50} values that were not significantly different for the 1 and 10 minute exposures in either sand or sandy loam soil but the 100 minute and overnight exposures provided significantly lower LC_{50} values (Table I). The CI for the Chlorfenapyr overnight exposure treatment in sand was not calculated because more than 90% mortality was recorded at all concentrations, as indicated by the slope (22.1) (Table I). None of the Fipronil treatments in sand were significantly different (Table I). In sandy loam soil LC_{50} values for Fipronil at 1 and 10 minute exposure times were not significantly different but these were significantly lower than the longer exposure times (100 minutes and overnight) (Table I). A CI for the Fipronil overnight exposure treatment in sand and sandy loam soil was not calculated because 100% mortality was recorded at two of the four concentrations tested. imidacloprid treatments in either sand or soil were not significantly different with less than 36% mortality regardless of concentration or exposure time (Tables I & II). At the time termites were removed from exposure to Imadichloprid, they appeared intoxicated (sluggish and unresponsive to stimuli such as opening the petri dish lid and prodding with forceps), yet most recovered and appeared normal (compared to the control group) after 24 h in the pesticide-free observation petri dish. No LC values were calculated for Indoxacarb because the range of concentrations tested in these bioassays provided an all or nothing response (Table II).

Probit analysis also indicated that there was a significant increase in the LC_{50} values comparing time of exposure between substrates—sand and sandy loam soil—for all of the termiticides tested (Table I). In every case, where slope values allow a statistically valid comparison, the LC_{50} values were significantly higher in sandy loam soil compared to the same exposure time on sand.

The corrected percent mean mortality data are provided by concentration and time of exposure for each termiticide in Table II. The 10 and 100 ppm concentrations are highlighted because labeled application rates for these chemistries are $\approx$ 50 ppm and field application would likely provide concentrations within this range. Fipronil in sandy loam soil was the only termiticide, regardless of substrate, to provide statistically significantly higher (ANOVA, LSD) mortality when the 10-minute exposure is compared to the 1-minute exposure (Table II). Within each termiticide the two shortest exposure times provided significantly lower mortality in sandy loam soil compared to sand except for the Imadichloprid at 10 ppm/10 minute exposure combination (4.0 ± 1.63 sand and 2.67 ± 1.18 sandy loam) (Table II).

Trends within each termiticide were not evident. Thiamethoxam on sand provided 100% mortality only at the 100 ppm concentration for the overnight exposure, although at 10 ppm in sand that percentage was 95% (Table II). The mean corrected percent mortality data with Thiamethoxam did not provide 100% mortality at either 10 or 100 ppm at any exposure time in sandy loam soil (Table II). Chlorfenapyr provided 100% mortality after overnight exposure in sand at 10 and 100 ppm and in sandy loam soil at 100 ppm (Table II). The sandy loam/100 ppm treatment provided 100% mortality although it was not

Table I. Comparison of LC $_{50}$ values, confidence intervals (CI) and slopes from the timed exposure bioassay by termiticide formulation, by exposure time and soil type

Time[1]	*LC $_{50}$*[2]	*CI*	*Slope ± SE*
IMIDACLOPRID			
		SAND	
1	11,290	2,844 to 119,867	0.45 ± 0.06
10	9,594	2,405 to 104,087	0.43 ± 0.06
100	10,950	2,032 to 278,547	0.33 ± 0.06
720	1,522	399 to 21,340	0.39 ± 0.08
		SANDY LOAM	
1	3×10^{12}	NA	0.11 ± 0.08
10	53×10^{6}	NA	0.25 ± 0.08
100	149×10^{12}	NA	0.03 ± 0.08
720	51,475	8,548 to 1,772,148	0.38 ± 0.07
FIPRONIL			
		SAND	
1	0.9	0.62 to 1.16	1.56 ± 0.20
10	0.9	0.71 to 1.13	2.35 ± 0.33
100	0.7	0.52 to 0.95	3.01 ± 0.57
720	0.9	NA	$20.84 \pm 1.46 \times 10^{5}$
		SANDY LOAM	
1	22	13 to 36	0.49 ± 0.06
10	17	11 to 24	0.69 ± 0.06
100	4	3 to 5	1.31 ± 0.12
720	2	NA	$9.07 \pm 1.75 \times 10^{4}$
THIAMETHOXAM			
		SAND	
1	35	25 to 49	0.84 ± 0.06
10	17	13 to 21	1.11 ± 0.07
100	5	4 to 7	1.35 ± 0.10
720	3	2 to 3	2.84 ± 0.36
		SANDY LOAM	
1	34×10^{6}	172,000 to 42×10^{6}	0.21 ± 0.07
10	1,870	990 to 4,516	0.67 ± 0.07
100	45	35 to 59	1.07 ± 0.07
720	12	9 to 15	1.19 ± 0.09

Continued on next page.

CHLORFENAPYR			
		SAND	
1	31	25 to 41	1.14 ± 0.07
10	23	19 to 30	1.23 ± 0.08
100	4	3 to 5	1.49 ± 0.11
720	0.9	NA	22.12 ± 1.45 × 10^5
		SANDY LOAM	
1	32,801	6,966 to 566,109	0.43 ± 0.07
10	2,996	1,269 to 11,076	0.51 ± 0.06
100	94	70 to 128	0.96 ± 0.07
720	10	9 to 13	1.87 ± 0.15

[1] Numbers represent minutes of exposure time.

[2] Value in ppm.

statistically different than the 100-minute exposure/100 ppm treatment (at 92% mortality) (Table II). Chlorfenapyr provided, in sandy loam soil, more than 50% mortality at 10 ppm for the overnight exposure but 100% for the same exposure time at 100 ppm (Table II). Fipronil provided 100% mortality at the overnight exposure at 10 and 100 ppm in either soil type (Table II). Fipronil was also the only termiticide to provide no statistical difference between any of the exposure times for both concentrations in sand. Yet, as with the other chemistries, in the sandy loam soil Fipronil showed significantly less mortality at the shorter exposure times, 1 and 10-minutes respectively (Table II). Imidacloprid regardless of concentration and exposure time provided less than 36% mean mortality in sand and < 11% mean mortality in sandy loam soil (Table II).

Table II. Comparison of mean corrected percent mortality from timed exposure bioassay by termiticide formulation, soil type and time of exposure at two concentrations

	SAND		*SANDY LOAM*	
Time[1]	*Mean*[2] ± *S.D.*		*Mean*[2] ± *S.D.*	
Imidacloprid	*10 ppm*			
1	10.0 ± 6.55	B[a]	2.67 ± 1.18	A
10	4.0 ± 1.63	B	2.67 ± 1.18	A
100	16.67 ± 7.28	BA	10.67 ± 6.72	A
720	36.25 ±18.5	A	7.7 ± 3.78	A
Imidacloprid	*100 ppm*			
1	14.67 ± 6.46	A	5.41 ± 1.68	A
10	25.33 ± 8.39	A	5.41 ± 2.38	A
100	22.91 ± 5.78	A	5.33 ± 2.74	A
720	35.97 ±15.45	A	2.0 ± 1.07	A

Thiamethoxam	*10 ppm*			
1	29.31 ± 9.25	C	9.33 ± 3	C
10	43.24 ± 8.7	BC	2.74 ± 1.22	C
100	68.3 ± 8.57	BA	22.89 ± 5.05	B
720	94.81 ± 2.32	A	36.39 ± 6.22	A
Thiamethoxam	*100 ppm*			
1	54.67 ±10.95	B	9.33 ± 3.3	C
10	75.22 ± 8.28	BA	14.81 ± 6.1	C
100	94.04 ± 3.29	A	62.59 ± 8.68	B
720	100 ± 0	A	86.15 ± 3.87	A
Chlorfenapyr	*10 ppm*			
1	22.31 ± 8.28	B	5.4 ± 2.38	B
10	32.67 ± 8.97	B	10.22 ± 3.74	B
100	86.67 ± 7.22	A	11.33 ± 4.67	B
720	100 ± 0	A	37.22 ± 8.27	A
Chlorfenapyr	*100 ppm*			
1	68.8 ± 9.82	B	4.0 ± 1.63	C
10	75.33 ±10.55	BA	12.0 ± 4.28	C
100	92 ± 5.54	BA	50.59 ±10.0	B
720	100 ± 0	A	100 ± 0	A
Fipronil	*10 ppm*			
1	95.04 ± 3.45	A	40.67 ± 9.43	C
10	98 ± 1.45	A	27.78 ± 7.26	C
100	99.3 ± 0.67	A	71.33 ± 9.65	B
720	100 ± 0	A	100 ± 0	A
Fipronil	*100 ppm*			
1	100 ± 0	A	46.67 ± 9.5	C
10	100 ± 0	A	71.15 ± 8.08	B
100	100 ± 0	A	96.52 ± 1.71	A
720	100 ± 0	A	100 ± 0	A
Indoxacarb	*10 ppm*			
1	100 ± 0	A	0	C
10	100 ± 0	A	0	C
100	100 ± 0	A	74 ± 5.3	B
720	100 ± 0	A	98 ± 1	A

Continued next page.

Indoxacarb		*100 ppm*		
1	100 ± 0	A	100 ± 0	A
10	100 ± 0	A	98 ± 1.7	A
100	100 ± 0	A	100 ± 0	A
720	100 ± 0	A	100 ± 0	A

[1] Time of exposure in minutes

[2] Mean corrected percent mortality followed by the standard error for that mean.

[a] Means followed by the same letter within the same column for each concentration indicate no significant difference (P=0.05).

Excavation Choice Bioassay

None of the termiticides tested provided 100% mortality after 21 days in bioassay at approximate labeled application concentrations (50-60 ppm) (Figure 1). Statistical separation of the various formulations was not performed because the purpose of these data was simply to illustrate that no termiticide eliminated all termites in a small bioassay arena arrangement over a 21 day period. All termiticide treatments, with the exception of Acetamiprid (n = 6), provided evidence that termites tunneled into the treated sand – chamber B (Figure 2). The controls, over the 21 days in bioassay, provided evidence of tunnels in all four soil arenas while none of the treatments provided similar data, indicating that even 'non-repellent' concentrations can, in a choice test design, have replicates that indicate 'repellence'. All termiticides were successful at 'protecting' the wood opposite the treated sand at the concentrations tested (Figure 3). These data indicate that all of the formulations tested would be effective at providing a barrier to termite infestation if applied to play sand at concentrations equivalent to 50-60 ppm.

Route of entry biosassay

The oral/nestmate treatment data did not differ from the oral/isolated treatment, regardless of termiticide, therefore those data are not provided in Table III. Fipronil provided the lowest values for either route of entry and was equally toxic by either route (Table III). Imidacloprid and Thiamethoxam were less toxic by the dermal/isolated compared to the oral route while Acetamiprid, Indoxacarb, and Chlorfenapyr were the opposite (Table III).

Fipronil displayed no difference in toxicity between the oral or dermal/isolated treatments and provided an additive affect when the dermal/nestmate or isolated regimes are compared, indicating that grooming activity provided an additional dose (Table III). The remaining termiticides provided data indicating that the most toxic route of entry (determined by the isolation treatment regime) for each formulation dominates toxicity if the behavior of the termite allows for the ability to groom similarly treated dermal-dosed nestmates. Chlorfenapyr, acetamiprid, and indoxacarb were less toxic by the oral route and the value for the dermal/nestmate was more than the dermal/isolated treatments, indicating that grooming activity probably increased the LD values in the dermal/nestmate treatment regime (Table III). Thiamethoxam and imidacloprid, despite a lower LD value for the oral route, provided no difference between the two dermal treatment regimes indicating that the intoxicated termites did not engage in grooming activity following application of the termiticide formulation (Table III).

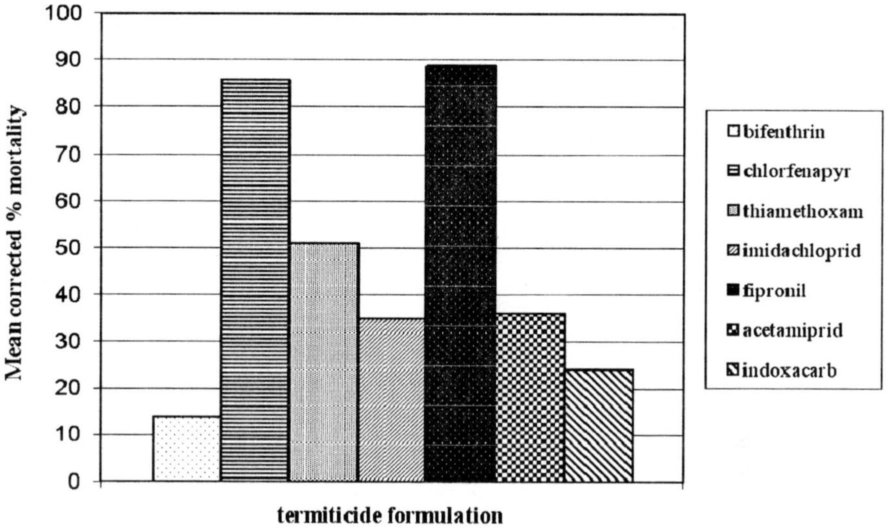

Figure 1. Mean corrected percent mortality from sand excavation choice bioassay at Day 21 by termiticide formulation.

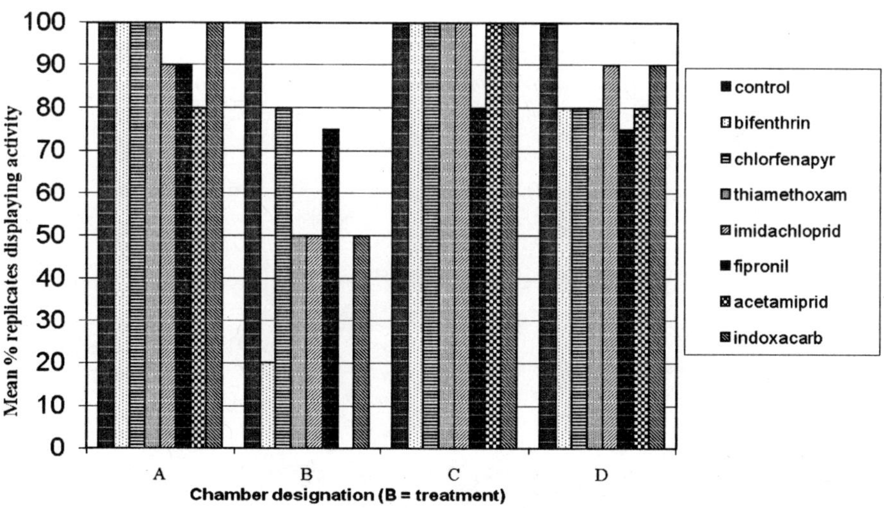

Figure 2. Mean percent of replicates from sand excavation bioassay that provided evidence of termite excavation by chamber and termiticide formulation at Day 21.

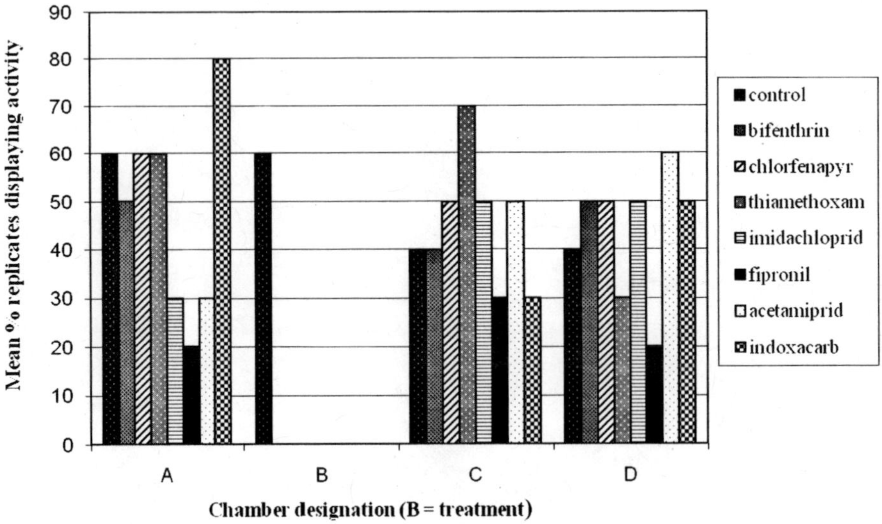

Figure 3. Mean percent of replicates from sand excavation bioassay that provided evidence of termite feeding on the wood by chamber and termiticide on Day 21.

Table III. Lethal Dose values at which 50% of the test animals died (LD50) for six selected termiticide formulations listed by active ingredient, route of entry (oral or dermal) treatment regime(dermal dose in isolation or with 9 similarly treated nestmates) in nanograms (ng) of active ingredient per gram of termite

Type of treatment regime	*LD50 VALUES (in ng AI per g of termite)* *Letters indicate significant difference based on CI overlap.*					
	Thiomethoxam	*imidacloprid*	*Acetamiprid*	*Fipronil*	*Indoxacarb*	*Chlorfenapyr*
Oral	238.8 A	931.7 A	277.34 A	22.57 A	303 A	11,152.9 A
Dermal nestmate	597 B	5,590 B	214.68 A	7.59 B	606 A	9,065.2 A
Dermal isolated	895.5 B	4,968.9 B	102.27 B	41.0 A	151.5 B	5,099.2 B

Timed running tests

All termiticides, within a respective chemistry, provided similar data trends comparing 1, 24 and 36-h after treatment and by 72-h a clear separation occurred. Therefore the 1- and 72-h data are provided in Figures 4 & 5. Termites dermal-dosed with imidacloprid were affected one hour after treatment as indicated by the high proportion of termites that did not run the full 6-cm 'test trail' and the longer time taken by those that did follow the trail (Figures 4 & 5).

The fipronil, indoxacarb, and chlorfenapyr treated termites were not affected one hour after treatment, as indicated by the high proportion of runners and fast running times (Figures 4 & 5). The Chlorfenapyr-treated termites were the exception in that they continued even up to 240-h (10 days after treatment) to provide results similar to the controls. Fipronil-treated termites, by 72-h, displayed 83% survivorship but the survivors were no longer responding to the ink; the few (17%) that did were sluggish with an average time of 22.3-s for the 6-cm. The survivorship of imidacloprid-treated termites was high (94%) at 72-h and a higher proportion (72%) of them ran compared to 1-h after treatment (33%). The lowest dose tested for acetamiprid and bifenthrin (0.02 ng/termite) provided immobilized subjects that did not survive 24 h so that data is not provided.

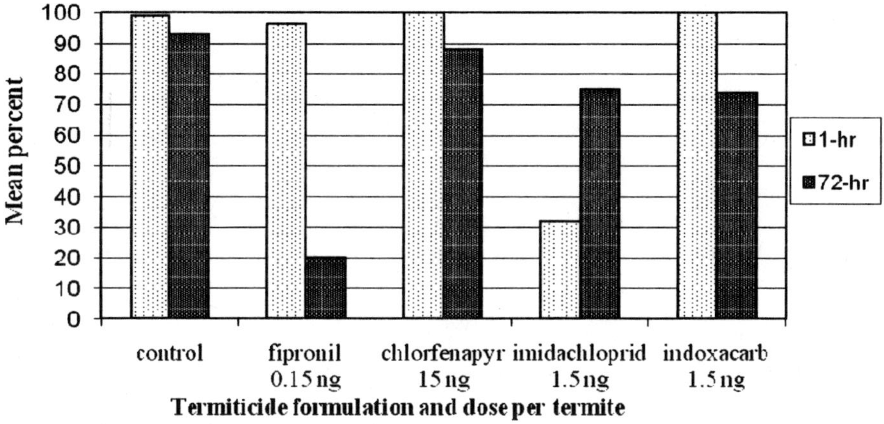

Figure 4. Mean percent topically-treated termites that ran along the ink pen line by termiticide formulation, dose in ng of active ingredient per termite and hours post-treatment.

Discussion

Congruence between laboratory screening bioassays and field efficacy is an important prerequisite for development of meaningful pesticide use patterns. This chapter attempts to provide data useful in modeling soil termiticide efficacy using a series of laboratory bioassays as a means of predicting field efficacy.

Attributes of termiticide chemistry that affect field efficacy of the end-use product would include persistence, bioavailability, solvent systems used in formulation, concentration, and active ingredient. Persistence is important for predicting the longevity of a soil barrier (*19*, *20*); The Chapter by Mulrooney, Wagner, and Gerard discusses appropriate methods for addressing this particular issue. Bioavailability is a complex issue and the comparisons between play sand and a sandy loam soil in the time exposure bioassay section of this chapter illustrate the importance of this phenomenon in understanding termiticide efficacy. The use of sand in termiticide bioassay has the advantage of repeatability between laboratories yet it should be plainly stated that field efficacy will most likely require higher concentrations for field results to mimic laboratory data (*22*). Solvent systems used in formulation can impact efficacy in bioassay as demonstrated by Smith and Rust (*23*) and Rust and Smith (*24*), who mention that proprietary information complicates testing formulation affects. A recent study demonstrated that formulated soil termiticides required contact to produce adverse affects and prompted the use of commercial products in the tests described in this chapter (*24*).

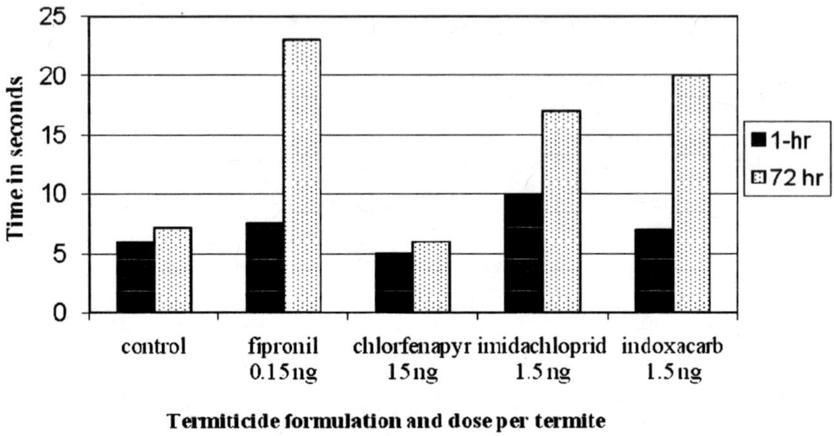

Figure 5. Time taken to run 6-cm by topically treated termites by termiticide formulation, dose in ng of active ingredient per termite and hours post-treatment.

Both concentration and the active ingredient are inexorably tied to their effect on the behavior of an individual termite. Intoxication of individuals, because of the collective decision making process employed by eusocial termites, must be considered within the context of what is known about termite behavior when interpreting bioassay data. It has long been recognized that insecticide concentration plays a significant role, as exemplified by studies attempting to find the 'repellent' limits in bioassay (*26, 133, 27, 28, 29*). However, concentration and route of entry are critical aspects of toxicity that have not been actively separated in previous studies. Dermal acquisition is generally assumed to be the major route of entry in the design of most termiticide bioassays (*13, 30, 28*). Saran and Rust (*29*) tested and supported the hypothesis that dermal contact is the major route of entry in bioassay involving termites exposed to treated substrates that cannot be excavated. The time exposure assays in this chapter, recent work by Green (*31*), and the fact that most of the published literature employ experimental designs that provide at least 1-hr exposure prior to examination of 'transfer' in bioassay (*13, 30, 28, 32*), illustrate that dermal acquisition is a slow process involving long-term (or multiple) exposure.

The mechanics, incidence and consistency of behaviors have an impact on designing bioassays with relevance to field efficacy. Two aspects of subterranean termite gallery construction are critical to interpreting bioassay data: the mechanics of the process and allocation of that task to individuals within a population. *R. flavipes* are known to manipulate soil particles with their mouthparts in the act of excavation (*33*). Therefore, termites involved in gallery construction following soil termiticide application or during foraging after application (curative or preventative scenario's respectively) would most likely be affected by an oral dose. Unfortunately there is a knowledge gap in how termites maintain galleries following excavation, but it can be assumed that this process also involves oral manipulation of gallery interior surface. Taken one step further these assumptions imply that termites not involved in gallery construction or maintenance (those using the gallery as an avenue of movement between feeding sites) would be the subjects that acquire a lethal dose through dermal contact.

It has been shown that not all worker termites are involved in gallery construction, indicating this task is performed by specific individuals (*33*); if they are killed or intoxicated, it can be assumed that communication of direction could eventually be lost to the main population. Applying the concept of swarm intelligence (*34, 35*) to termiticide bioassay would suggest that tests in perti dishes or tubes oversimplify the process used by intact termite field populations when confronted with a soil termiticide application. The system employed by subterranean termites to communicate traffic flow within the network of galleries connecting different feeding sites is unknown and without more knowledge a choice bioassay offers a better approximation of field events compared to a no-choice system.

This chapter attempts to illustrate relevant aspects of termiticides that require consideration when designing termiticide bioassay. The first is that the term "non-repellent" is not the purview of any particular class of chemistries but is a matter of concentration. The four-way choice bioassay data reported from these trials clearly demonstrated that all of the termiticides tested do not provide 100% mortality in small choice-test arenas after 21 days of exposure (Figure 1). Depending on concentration, all of the chemistries tested, including a 'repellent' formulation (bifenthrin), provided data where tunneling and mortality was limited (Figures 1-3). These data demonstrate the subjective nature of the non-repellent label and that this moniker should be used with a caveat to concentration. The second is bioavailability. It has been demonstrated in other studies that bioavailability is important for predicting the efficacy of a particular termiticide (*36, 37, 6, 38, 15, 22*). All of the termiticides tested in the studies outlined in this chapter, with the exception of imidacloprid, provided decreased termite mortality when comparing the same treatment between soil types (Table II). Bioavailability is a complex phenomenon that plays a role in termiticide field efficacy, especially at low concentrations. However, it is clear that one should use caution when extrapolating laboratory results obtained from bioassay in sand – especially in regard to field-use recommendations.

The third aspect of soil termiticide efficacy demonstrated by these tests involves route of entry. The dose-mortality assays, reported herein, clearly demonstrate the most toxic route of entry varies for each of the chemistries tested (Table III). Termites can obtain a lethal dermal dose one of two ways: body-to-body contact with a dermal-contaminated termite or a contaminated substrate (gallery or soil). The timed exposure data (Tables I & II) indicate that in the field repeated exposures would be required to produce significant mortality by the dermal route of entry. Following a label application there would be, under ideal conditions, a 16-cm zone of treated soil that termites could traverse - once a gallery is constructed - in less than one minute. The one-minute exposure mortality data for sandy loam soils did not produce sufficient mortality by the dermal route of exposure to justify the attrition hypothesis (Table II). Shelton and Grace (*30*) demonstrated that, following exposures times ranging from 3-24 h on soil containing 1 ppm of fipronil or imidacloprid, transfer of a lethal dose was unpredictable and provided no more than 26% mortality in unexposed 'recipients'. Saran and Rust (*29*) exposed termites to treated sand for 1-hr and found limited dermal uptake of ^{14}C fipronil with most occurring by constant exposure for 24-h. The lessons from these and other studies is that the most likely route of entry following field application of a termiticide will be an oral dose through soil manipulation during gallery construction (*33*) and gallery maintenance - not the dermal route.

The potential that grooming could play to provide an oral dose for termiticide transfer has been demonstrated by Myles (*39*). Grooming has been documented as the most consistently performed behavior displayed by worker termites (*41*) and Whitman (*40*) found it occupied 16% of the active time of the average worker. These data indicate the potential grooming would have as a mechanism of oral dose transfer, but the dermal acquisition data would arguably relegate the grooming-oral route to a minor role unless the active ingredient provides a favorable oral/dermal toxicity profile (Table III). Another behavior that could contribute to transfer by the oral route would be food exchange, yet most studies of trophallaxis do not separate the potential modes - stomodeal or proctodeal (*41, 42, 43, 44, 45*). The trophallaxis oral route would be minimal because the amount of labeled food transferred is estimated from the literature to be below 15% on any given day (*44, 45*). Whitman (*40*) corroborates those data by describing stomodeal exchanges involved only food being chewed, eliminating it as a source of soil termiticide transfer, and proctodeal exchanges accounted for less than 1/3 of the food intake for the average worker over the course of several days. Cannibalism is the last potential route of oral dose acquisition but it has not been purposefully studied and seems an unlikely major contributor to any model of termiticide transfer given the reports of cadaver burying behavior in termites (*46*).

The fourth aspect of soil termiticide efficacy involves an understanding of termite behavior relative to collective decision making within termite populations. The importance of the decision making process is illustrated by data that incorporate distance in the experimental design whether laboratory bioassay (*47, 29*) or field studies (*48*) that do not provide evidence of transfer beyond a few meters. Any termite that is contaminated by contact with a termiticide represents a potential toxicant delivery system to other parts of the network of galleries and feeding sites maintained by a population of termites. The contaminated 'agent of transfer' must however be behaviorally unaffected long enough to exit the area of exposure. The timed running tests provided in the chapter and the work of Saran and Rust (*29*) represent a starting point toward understanding the potential a particular termiticide has toward movement by contaminated termites as well as the range of doses that would allow transfer in the field. The data to date indicate that the dose response range is not only specific to a certain chemistry (and route of entry) but for most termiticides is represented by a small range of concentrations. The effective dose range required for any specific termiticide that would allow movement of 'lethal donors' from the point of insecticide application to a significant number of termites in a field population needs to be examined in more detail. It would appear, however, even with this level of detailed understanding that field application of a liquid soil termiticide could not, given the plethora of soil types and conditions at any single treatment site, provide consistent realization of the attrition hypothesis. Consistently attaining this range of concentrations in a field application would be impossible and combined with the impact of bioavailability, route of entry, population pressure, and persistence makes accurate predictions problematic.

In the field, aversion also may play a role in the efficacy of termiticides. Thorne and Breisch (*49*) determined that termites exposed to a sublethal dose of imidacloprid did not 'learn' to avoid treated soil. However, if termites get 'sick' in certain galleries those routes may receive less traffic and the chemical messages indicating a "path-to-follow" may deteriorate thereby mimicking aversion. Similarly, termites excavating into a soil termiticide treatment could die quickly, for example at the active end of gallery construction, and the chemical signal to travel down that path - at the fork in the system where termites decide which gallery to traverse - would be lost and activity redirected to another portion of the gallery system. The end result of collective decision making in termite gallery traffic patterns could result in structural protection at termiticide concentrations that kill termites (the traditional non-repellent concept) but do not impact termite populations (as in the traditional repellent paradigm). Repeated attempts to dig through treated soil could, however, reduce the number of termites, assuming all termites are involved in gallery construction, with structural protection achieved in a manner consistent with the attrition hypothesis although transfer played no role.

Summary

Advances in our understanding of termite biology and the new chemistries registered since 1982 call for a paradigm shift in bioassay design for testing termiticide efficacy. It is clear from the series of bioassays reported in this chapter, in addition to other studies (*47, 48, 32, 29*), that the attrition model of termiticide efficacy is unlikely to be consistently realized in the field using soil-based application of the formulations tested in this study. The mechanics of gallery construction and attendant oral dose, the realization that gallery construction is conducted by specific individuals combined with the assumption that traffic through the gallery system is dictated by collective decision making begs a reevaluation of soil termiticide efficacy. I propose that soil termiticides act as a preventative barrier following treatment because termites allocated to exploring for new food resources (termites involved in gallery construction, foragers) are killed (slowly or otherwise) and the gallery leading to a soil treatment goes unexploited/unused by the remaining population whose traffic flow is directed elsewhere. Soil termiticides work in eliminating active structural infestations in much the same way by redirecting traffic in the soil in combination with killing termites 'trapped' in the structure through desiccation or contact with the treated soil.

Soil termiticides must be evaluated in conjunction with an understanding of subterranean termite behavior, as illustrated over 20 years ago by Su *et al.* (*4*), but advances in our understanding of termite behavior begs prudent interpretation of single-design bioassay data. Information from a bioassay series can be applied to what is known about termite foraging and colonization activities to formulate a hierarchy of outcomes based on the attrition or barrier hypotheses of soil termiticide efficacy and can be used to design active ingredients and application methodologies that would optimize transfer to realize widespread attrition. Determination of the oral and dermal toxicities combined with measures of the time-frame for toxicity-related behavioral changes could be used to predict useful insecticide candidates for realization of the attrition model However, actual field efficacy will always be subject to the vicarious conditions present at an individual treatment location, which may never be anticipated with certainty, highlighting the importance of attention to the details of application by the end user.

Acknowledgements

I thank Mark Yates, Lisa Stabler and Sam Wise for their invaluable technical assistance in managing the conduct of the bioassays and analyzing results. Two unknown reviewers must be thanked for their useful comments and Bethany Farrey for her technical proofreading skills that combined to make this a more readable piece. Lastly, I want to acknowledge the contributions of FMC, BASF, DuPont, Dow Agrosciences, and Syngenta for their partial funding of the various research projects that lead to the data used in this chapter.

References

1. Forschler, B. T. *NPMA Research Report on Subterranean Termites*; NPMA: Dunn Loring, VA, 1998; pp. 31-51.
2. Potter, M. F. *Termites. In Mallis Handbook of Pest Control*, 8th ed.; Mallis Handbook & Technical Training Co.: Cleveland, OH, 1997; pp. 233-333.
3. Ebeling, W.; Pence, R. J. *J. Econ. Entomol.* **1958**, *51*, 207-211.
4. Su, N.-Y.; Tamashiro, M.; Yates, J. R.; Haverty, M. I. *J. Econ. Entomol.* **1982**, *75*, 188-193.
5. Su, N.-Y.; Wheeler, G. S.; Scheffrahn, R. H. *J. Econ. Entomol.* **1995**, *88*, 1690-1694.
6. Gold, R. E.; Howell Jr., H. N.; Pawson, B. M.; Wright, M. S.; Lutz, J. C. *Sociobiology*. **1996**, *28*, 337-364.

7. Kard, B. M. *Pest Control*. January, 2001, p 30-33, 73.
8. Jones, S. C. *Pest Manage.* February, 1989, p 16-18.
9. Su, N.-Y.; Scheffrahn, R. H.; Ban, P. M. *J. Econ. Entomol.* **1993**, *90*, 503-509.
10. Forschler, B. T. *J. Entomol. Sci.* **1994**, *29*, 43-54.
11. Kuriachan, I.; Gold, R. E. *Sociobiology*. **1998**, *32*, 151-166.
12. Potter, M. F.; Hillary, A. E. *Sociobiology*. **2002**, *39*, 373-405.
13. Ibrahim, S. A.; Henderson, G.; Fei, H. *J. Econ. Entomol.* **2003**, *96*, 461-467.
14. LaFage, J. P.; Su, N.-Y.; Jones, M. J. *Sociobiology*. **1983**, *7*, 305-310.
15. Forschler, B. T.; Townsend M. L. *J. Econ. Entomol.* **1996**, *89*, 678-681.
16. Scheffrahn, R. H.; Su, N.-Y. *Fla. Entomol.* **1994**, *77*, 460-474.
17. Chen, J.; Henderson, G.; Laine, R.A. *J. Entomol. Sci.* **1998**, *33*, 97- 105.
18. SAS Institute. *SAS Users Guide*, version 8.2. SAS Institute: Cary, NC, 1999.
19. Grace, J. K.; Yates, J. R.; Tamashiro, M.; Yamamoto, R. T. *J. Econ. Entomol.* **1993**, *86*, 761-766.
20. Saran, R. K. M.S. Thesis, University of Nebraska, Lincoln, NE, 2001.
21. Mulrooney, J. E.; Wagner, T. L.; Gerard, P. D. Fipronil: Toxicity to Subterranean Termites and Dissipation in Soils. In *Pesticides in Household, Structural and Residential Pest Management*; Peterson, C. J., Stout, D., II, Eds.; American Chemical Society: Washington, DC, 2009; 107-123.
22. Osbrink, W. L. A.; Lax, A. R. *J. Econ. Entomol.* **2002**, *95*, 989-1000.
23. Smith, J. L.; Rust, M. K. *J. Econ. Entomol.* **1991**, *84*, 181-184.
24. Rust, M. L.; Smith, J. L. *J. Econ. Entomol.* **1993**, *83*, 1131-1135.
25. Delgarde, S.; Rouland-Lefevre, C. *J. Econ. Entomol.* **2002**, *95*, 531-536.
26. Gahlhoff, J. E.; Koehler, P. G. *J. Econ. Entomol.* **2001**, *94*, 486-491.
27. Remmen, L. N.; Su, N.-Y. *J. Econ. Entomol.* **2005**, *98*, 906-910.
28. Hu, X.P. *J. Econ. Entomol.* **2005**, *98*, 509-517.
29. Saran, R. K.; Rust, M. K. *J. Econ. Entomol.* **2007**, *100*, 495-508.
30. Shelton, T. G.; Grace, J. K. *J. Econ. Entomol.* **2003**, *96*, 456-460.
31. Green, J. M.S. thesis, Purdue University, West Lafayette, IN, 2007.
32. Shelton, T. G.; Mulrooney, J. E.; Wagner, T. L. *J. Econ. Entomol.* **2006**, *99*, 886-892.
33. Whitman, J. G.; Forschler, B. T. *Ann. Entomol. Soc. Am.* **2007**, *100*, 763-771.
34. Condrat, L.; Roper, T. J. *Trends in Ecol. and Evol.* **2005**, *20*, 449-456.
35. Bonabeau, E.; Dorigo, M.; Theraulaz, G. *Swarm intelligence: from natural to artificial systems.* Oxford University Press: New York, NY, pp 275.
36. Harris, C. R. *Annu. Rev. Entomol.* **1972**, *17*, 177-198.
37. Smith, J. L.; Rust, M. K. *J. Econ. Entomol.* **1993**, *86*, 53-60.
38. Gold, R. E.; Howell, Jr., N. H.; Pawson, B. M.; Wright, M. S.; Lutz, J.C. In *Proceedings, 2nd International Conference on Insect Pests in the Urban Environment, 7-10 July 1996*; K. B. Wildey, Ed.; Heriot-Watt University: Edinburgh, Scotland, 1996, pp 467-484.
39. Myles. T. G. *Sociobiology*. **1996**, *28*, 373-400.
40. Whitman, J. G. M.S. Thesis, University of Georgia, Athens, GA, 2006.
41. Rosengaus, R. B.; Traniello, J. F. A.; Levy, C. K. *J. Appl. Enomol.* **1986**, *101*, 287-294.
42. Cabrera, B. J.; Rust, M. K. *Insectes Soc.* **1999**, *46*, 244-249.

43. Sheets, J. S.; Karr, L. L.; Dripps, J. E. *J. Econ. Entomol.* **2000**, *93*, 871-877.
44. Suarez, M. E.; Thorne, B. L. *Ann. Entomol. Soc. Am.* **2000**, *93*, 145-155.
45. Saran, R. K.; Rust, M. K. *J. Econ. Entomol.* **2005**, *98*, 1284-1293.
46. Roy, H. E.; Steinkraus, D. C.; Eilenberg, J.; Hajek, A. E.; Pell, J. K. *Ann. Review Entomol.* **2006**, *51*, 331-3357.
47. Su, N.-Y. *J. Econ. Entomol.* **2005**, *98*, 2143-2152.
48. Osbrink, W. L. A.; Cornelius, M. L.; Lax, A. R. *J. Econ. Entomol.* **2005**, *98*, 2160-2168.
49. Thorne, B. L.; Breisch, N. L. *J. Econ. Entomol.* **2001**, *94*, 492-498.

Chapter 6

Colony differences in termiticide transfer studies, a role for behavior?

Thomas G. Shelton

USDA Forest Service; Insects, Diseases, and Invasive Plants; Starkville MS

Abstract. Donor-recipient termiticide transfer laboratory tests were performed by using destructive sampling with two delayed-action non-repellent (DANR) termiticides against each of three colonies of *Reticulitermes flavipes* (Kollar). Two of the three colonies showed no response to indoxacarb, but all three showed a response to chlorantraniliprole. These results indicate that behavioral variation among colonies is not likely responsible for the variability in recipient mortality among colonies noted in transfer studies in the literature. Donor mortality with these compounds and colonies suggests that variable physiological susceptibility of individual colonies to certain compounds may be more important than variations in behavior.

Introduction

Over the past several years, laboratory studies on the movement of termiticides among termites have been reported in the pest control (*1, 2*) and scientific literature (*3 - 10*). These reports have generally found that delayed-action non-repellent (DANR) termiticides are capable of movement among termites, while more traditional repellent compounds do not move among those termites directly exposed (*11*).

Recent studies at the University of California, Riverside (*8, 10*) have indicated that for some compounds, lethal termiticide transfer can only happen via primary transfer. In other words individuals exposed to a toxicant (donors) may pass it to naïve termites (recipients), but those recipients do not become secondary donors themselves, mainly due to the limited amount of toxicant available from the original donor (*8*). There may also be a location component to this situation: termites exposed to treated soil (donors) would have the toxicant coating their cuticle, whereas recipients picking up the toxicant via grooming and/or trophallaxis would ingest the materials, thus making them unavailable for movement via grooming. However, transfer via trophallaxis, or proctodeal feeding could still occur.

Toxicant transfer has only been documented in the laboratory, although anecdotal evidence from field studies has been used to support its occurrence in the field (*1, 12*). Without direct evidence of the effects of termiticide transfer on field populations there are serious doubts about the importance of transfer in real-world control situations. This is perhaps best demonstrated in a field study by Osbrink *et al.* (*13*), in which soil application of imidacloprid a DANR termiticide previously shown to transfer among *Coptotermes formosanus* Shiraki individuals in the laboratory (*6*), did not lead to population suppression of *C. formosanus* consistent with what the authors termed a liquid-bait model. For the purposes of this chapter, the idea of "termiticide transfer" is limited to mortality induced by the movement of soil-applied chemicals and not termiticidal baits. Bait products are designed to be consumed by termites and spread throughout colonies via social interaction, whereas the transfer discussed in this chapter consists of movement of chemicals from the soil (not necessarily consumed) to exposed and eventually to naïve termites (almost an "accident" in terms of product design). For information on the movement of a bait toxicant (hexaflumuron), the reader is referred to Sheets *et al.* (*14*). Unfortunately the most useful tool for examining transfer is the use of radiolabelled termiticides, which are unikely to receive approval for use in field experiments. Transfer of termiticides also may not have large effects on foraging populations as mortality in these laboratory studies can be quite low (*6, 9*). To some extent these problems have made termiticide transfer into more of an academic curiosity, a phenomenon to be studied certainly, but not to be relied upon for termite control.

Because there are problems with field observations of transfer, although such data are sorely needed, examinations of transfer are best suited to laboratory work. Some authors have chosen to examine termite colony origin in relation to termiticide transfer (*4, 6, 11, 9*). Differences in recipient mortality among colonies exposed to the same concentrations of single toxicants could have a number of possible explanations (*3, 6*). In previous papers, termite body mass did not predict susceptibility to toxicant transfer (*i.e.*, low body mass was not associated with increased susceptibility, nor vice versa). Dosage may also be an issue with studies where termites are left to walk across treated surfaces, as there is little guarantee of the consistency of the dose received (compared with topical applications). Topical applications, however, unless based upon known concentrations picked up by termites interacting with treated soil, are themselves rather arbitrary. It has been suggested that behavioral differences among colonies may lead to such differences in horizontal transmission mortality (*15, 14, 3, 6*), keeping in mind that recipient mortality depends upon the movement of toxicants from donors to recipients over the course of the study. It is easy to imagine how the rates of behaviors such as grooming or trophallaxis could influence the rate at which toxicants are passed among individuals, with particularly low rates possibly leading to lack of transfer altogether.

How can the impact of intercolony behavioral differences be tested? One way of testing this hypothesis involves making a simple assumption that these behavioral rates are consistent within colonies, but not necessarily among colonies. It should be kept in mind that it would be unreasonable to assume that some colonies simply do not engage in all social behaviors (assuming Occam's razor has taken out any unnecessary behaviors via evolution), however different colonies might very easily vary in their rates of conducting these behaviors.

Work with lower termites (Termopsidae, Kalotermitidae) has indicated that all non-larval (*i.e.* third instar and above) "workers" or pseudergates seem to take on equivalent roles (engage in the same sets of behaviors) within colonies (*16, 17*). There is a little evidence here for temporal division of labor; Howse (*16*) found that within the pseudergate caste of *Zootermopsis nevadensis* (Hagen), first and second instars exhibited little to no behaviors aside from trophallaxis, and that the amount of time spent in trophallaxis for all stages had a weak inverse correlation with time spent in other activites such as building or digging. Although no inference testing was done, some of the non-trophallaxis behaviors (building and oscillatory movements) increased with pseudergate instar, but others remained steady (digging). In general the sixth instar performed most of the colony "work" activities (*16*). The concepts suggested by Howse (*16*) were verified in papers by Crosland and colleagues (*18, 19*) with *Reticulitermes fukienensis* Light, indicating that all pseudergates perform the same behaviors, but the rates of those behaviors varied with age class. *Reticulitermes flavipes* (the model insect for this paper) often includes larval termites in the foraging populations (*20*), however most laboratory studies with this species do not include larval stages or nymphs (wing-pad bearing pre-alates), unless otherwise noted. Additionally, one might expect the age class distribution of a colony to be somewhat steady at any given point in time, thus as long as only pseudergates are counted (no larvae, nymphs) without bias into groups for an experiment, those groups (for a given colony) should be fairly similar in terms of age class distribution. Thus, if there were a temporal division of labor in *R. flavipes*, it would be unlikely to invalidate the assumption due to experimental methodology. Testing the hypothesis above requires looking for intracolony differences in recipient mortality in transfer studies with toxicants that have been previously shown to transfer among termites, using the same colonies (and thereby keeping the rates constant).

Saran and Rust (*10*) provide some insight into what behaviors might be most important in toxicant transfer. They examined movement of fipronil among *R. hesperus* Banks whose mouthparts had been sealed with glue. Saran and Rust (*10*) conclude that movement of the toxicant did not rely on trophallaxis at all, only body contact. However, with sealed mouthparts, other behaviors, such as grooming, would also be impeded, and this was attributed to the movement of imidacloprid in Tomalski and Vargo's study (*2*). These researchers also found that corpses of termites killed with fipronil were toxic enough to kill recipients, depending on concentration. Their data indicate trophallaxis and proctodeal feeding are not necessary for transfer with *R. hesperus*, although it is still possible that these behaviors (in addition to grooming) might accelerate the movement of termiticides.

From the arguments above, it can be assumed that any given colony will have constant rates of trophallaxis/grooming that might allow for transfer to occur. The hypothesis is that these behaviors have some influence on the rate of recipient mortality due to insecticide transfer and are responsible for the differences among colony mortalities noted in previous studies (alternate hypothesis; H_a). To make this a testable hypothesis, we need something to examine: two pesticides capable of being transferred as toxicants against a single colony in a simple donor-recipient transfer laboratory study (with replications being the testing of new colonies). If the hypothesis stated is correct, then both compounds will produce similar results for each individual colony. Finding non-relative differences (i.e. not just a small difference in percent mortality, but rather that one compound has an effect and the other does not) in each colony's response to individual compounds would not support the alternative hypothesis of behavior influencing recipient mortality and would indicate that some other factor is responsible for the differences noted in previous studies (null hypothesis; H_0).

The following study investigates this idea using two new termiticides produced by E. I. DuPont de Nemours, Inc. The first is an oxadiazine compound known as indoxacarb, and the second is chlorantraniliprole (class anthranilic diamide). The literature provides some information on transfer with indoxacarb (*7*) against another subterranean termite species, *C. formosanus*. Both compounds are capable of being transferred by *Reticulitermes flavipes* (Kollar) using 5% of the test population as donors exposed to 100 ppm of toxicant treated sand, as determined in preliminary studies.

Methods and Materials

Termites. Termites were collected by removing infested logs (cut into manageable sections) from active termite colony sites in the John W. Starr forest (maintained by Mississippi State University), the Noxubee National Wildlife Refuge (maintained by the U.S. Fish and Wildlife Service), and the campus of the USDA Forest Service facility in Starkville MS (all termite colonies were collected within 15 miles of Starkville). Log sections were placed into 30 gal. (114 L) metal trash cans, and returned to the laboratory, where cans were kept under ambient conditions (~24 C) until use. Termites were identified from morphological soldier characters by using the key of Hostettler *et al.* (*21*).

The studies were simple donor-recipient mortality studies run for two weeks. However, to get a more detailed view of mortality over this time period, the tests were run by using destructive sampling, with 6 replicates from each treatment (controls and 100 ppm pesticide) broken down and surviving termites counted on every other day during the test period (2, 4, 6, 8, 10, 12, and 14 days after treatment). Due to the number of replicates needed for this sampling method, only two treatments per compound were included (a distilled water-only control, and a 100 ppm pesticide treatment). For each colony, tests for each compound were run in separate incubators (25 ± 1 °C; ~75% R.H.) with separate control groups for each compound (*i.e.*, indoxacarb replicates + indoxacarb controls in incubator 1, chlorantraniliprole + control replicates in incubator 2). For each colony, 168 experimental units (jars) were necessary. Data sets were collected for both compounds using each colony.

With the exception of the destructive sampling, the methods for each study were a modification of those previously published (*9*). Donors were stained by feeding them filter papers (Whatman #2, Whatman International Ltd., Maidstone, United Kingdom) stained with Sudan Red 7B (0.5% wt./wt.; Sigma-Aldrich co., St. Louis, MO; *22*) for one week prior to the start of the test. Staining took place in Petri dishes (9 cm dia.) provided with two stained filter papers, moistened with 1 ml of distilled water each, containing 200-250 termites (mixed caste) and incubated at 25 ± 1 °C, ~75% R.H in an unlit incubator. Arenas were standard 8 cm diameter × 10 cm tall screw top plastic Quorpak jars, filled with 150 g of silica sand (Fisherbrand; Fisher Scientific, Pittsburgh, PA), and moistened with 27 ml of distilled water. On the test initiation day, recipient termites were counted fresh from the cans into groups of 95 workers only, and one group was placed into each jar. Donor termites were counted into six groups of 100 workers and placed in petri dishes containing 25 g treated sand (three dishes per treatment; either water only or 100 ppm wt./wt. of pesticide/sand), which was provided with 6 ml of distilled water 3 hrs prior to adding termites (to allow for evaporation). Donor groups spent 1 hr on the treated sand (consistent with previous papers on this subject: *6, 11, 9*) before being moved to clean petri dishes containing only a single dry filter paper for 30 min (this allowed any sand attached to the donors to dislodge). Finally, donors were placed into jars according to treatment, at a rate of 5 donors per jar. On breakdown days (described above) jars were emptied onto plastic trays and surviving donor (as stained individuals) and recipient workers were counted and recorded.

Statistically, each compound + control grouping (per colony) was considered separately, with percentage recipient mortality transformed by the arcsine of the square root and subjected to a general linear model procedure (GLM, *23*). Concentration of pesticide, day of test, and their interaction were investigated for influence on recipient mortality. Of these, the most important measure is that of concentration, which indicates whether transfer of the pesticide led to mortality of the recipients. Certainly, transfer which does not lead to recipient mortality cannot be measured using these methods, but sub-lethal movement of pesticides is not the metric being examined here.

Results

Figures 1 and 2 illustrate mean ± SEM percentage mortality for donors (Figure 1) and recipients (Figure 2) by colony and compound for these studies. Details of each colony's response to both compounds are given separately below.

Colony 1. Donor mortality shows a trend with chlorantraniliprole increasing donor mortality until roughly day 6, then leveling off (Figure 1). Indoxacarb treated donors begin showing mortality at day 10 then leveling off below 40% (Figure 1). Mortality of donors from colony 1 was significantly influenced by concentration of both insecticides (chlorantraniliprole: dF = 1, 83;

F = 289.69; P < 0.0001; indoxacarb: dF = 1, 83; F = 12.12; P = 0.0009). Concentration of chlorantraniliprole significantly influenced recipient mortality of colony 1 workers compared to controls during this study (dF = 1, 83; F = 58.29; P < 0.0001), but concentration of indoxacarb did not significantly influence recipient mortality (dF = 1, 83; F = 3.18; P = 0.0792) in comparision to untreated controls. Recipient mortality increased until day 6 for chlorantraniliprole treated replicates, and leveled off after that point (Figure 2), but did not increase with indoxacarb over the 14 day test (Figure 2) for colony 1 termites. Recipient mortality with both compounds was not significantly influenced by day of test (chlorantraniliprole: dF = 6, 83; F = 1.85; P = 0.1020; indoxacarb: dF = 6, 83; F = 2.15; P = 0.0584). The interaction of concentration by day significantly influenced recipient mortality only for indoxacarb for colony 1 workers (dF = 6, 83; F = 2.31; P = 0.0429; chlorantraniliprole: dF = 6, 83; F = 1.60; P = 0.1594).

Colony 2. During breakdown of replicates for this colony, some individuals in five replicates were noted to have a bright red coloration commonly associated with *Serratia* sp. infection (note that Sudan Red staining results in a much deeper red color). These termites only showed up in five replicates of the chlorantraniliprole treatment (two on day six, one on day eight, and two on day 12). These replicates were left out of the analysis, as well as Figures 1 and 2.

Donor mortality for colony 2 is quite similar to the response of colony 1, with chlorantraniliprole donor mortality leveling off by day 6 (reaching 100% by day 10; Figure 1). For indoxacarb, donor mortality is slightly increased above that of control recipients, eventually overlapping on day 14 (Figure 1). As with colony 1, colony 2 donor mortality was significantly influenced by concentration for both compounds (chlorantraniliprole: dF = 1, 78; F = 382.85; P < 0.0001; indoxacarb: dF = 1, 78; F = 10.68; P = 0.0017). Concentration of chlorantraniliprole significantly influenced recipient mortality of colony 2 workers (dF = 1, 78; F = 133.93; P < 0.0001). Indoxacarb did not significantly influence recipient mortality of colony 2 workers (dF = 1, 78; F = 2.08; P = 0.1539), also similar to the results obtained with colony 1. Chlorantraniliprole treatment recipients never seemed to reach a plateau for colony 2 termites, although the slope of the data changes at around day 10 (Figure 2). For indoxacarb, recipient mortality essentially mimics control recipient mortality for the entire duration (Figure 2). For colony 2 workers, day of test significantly influenced recipient mortality for both compounds tested (chlorantraniliprole: dF = 6, 78; F = 6.63; P < 0.0001; indoxacarb: dF = 6, 78; F = 16.94; P < 0.0001). For colony 2 workers the interaction of day and concentration significantly influenced recipient mortality only for chlorantraniliprole (dF = 6, 78; F = 3.06; P = 0.0106; indoxacarb: dF = 6, 78; F = 0.28; P = 0.9432).

Colony 3. Donor mortality (Figure 1) follows the same path for both indoxacarb and chlorantraniliprole for colony 3 termites, in that both reach a maximum (100% mean donor mortality) on day 6, which holds for the remainder of the study. As with both other colonies, colony 3 donor mortality was significantly influenced by concentration for both compounds (chlorantraniliprole: dF = 1, 83; F = 332.42; P< 0.0001; indoxacarb: dF = 1, 83; F = 511.40; P < 0.0001). As with colonies 1 and 2, chlorantraniliprole concentration significantly influenced recipient mortality of colony 3 workers (dF = 1, 83; F = 85.43; $P < 0.0001$). Colony 3 responded differently to indoxacarb than colonies 1 and 2, in that indoxacarb concentration significantly influenced recipient mortality of colony 3 workers (dF = 1, 83; F 164.39; $P < 0.0001$). Figure 2 indicates that in indoxacarb treatments, recipient mortality increases until roughly day 6, when it plateaus for several days, increasing again on the final day (day 14). Recipient mortality in chlorantraniliprole treatments spikes fairly early (day 4) with colony 3 termites, and then falls to a plateau for the remainder of the study (Figure 2). For colony 3 workers, day of test with both compounds significantly influenced recipient mortality (chlorantraniliprole: dF = 6, 83; F = 2.80; $P = 0.0167$; indoxacarb: dF = 6, 83; F = 9.45; $P < 0.0001$). As with colony 1 workers, colony 3 worker recipient mortality was significantly influenced only by indoxacarb in the day by concentration interaction (dF = 6, 83; F = 2.49; $P = 0.0306$; chlorantraniliprole: dF = 6, 83; F = 1.24; $P = 0.2948$).

Discussion

Figure 2 illustrates recipient mortality during these studies, with indoxacarb data in the left column and chlorantraniliprole data in the right column. By viewing each colony's response to these compounds individually, it is obvious that colony 1 and colony 2 did not respond in a similar manner to both compounds. The GLM analysis of these data confirm that both colonies' recipient mortality was significantly influenced by exposure to donors treated with chlorantraniliprole only. Colony 3 recipient termite mortality was significantly influenced by both indoxacarb and chlorantraniliprole treated donors. There are two possibilities suggested by these results: a) the initial assumption regarding the consistency of behavioral rates within any given colony is not correct for *R. flavipes* workers, or b) intercolonial variability in recipient mortality in transfer studies previously reported is not due to variations in behavioral rates among colonies.

While the behavioral rate consistency assumption seems plausible, there have been no attempts to determine the rate differences (if any) among the workers within these colonies. Evidence from other lower termites suggests that this is not an unreasonable assumption. Studies with *Z. nevadensis, R. fukienensis,* and *Kalotermes flavicollis* (Fabricius) indicate that workers engage in similar behavioral capacities within fairly broad age groups (*16, 18, 19, 17*). In other words, termites beyond the 2nd instar are engaged in similar activites as other workers up to the pre-alate nymph stage (*18, 19, 17*). It should be noted that this assumption would certainly be invalid for some social Hymenoptera (*24*). The possibility of temporal division of labor as suggested earlier remains, although the age class distribution of groups counted from these colonies should have been similar within each colony as discussed in the introduction. Variations in the performance of behaviors have been noted between colonies for behaviors such as tunnel building (*25*) and agonism (*26 - 28*). However, variations in rates of behaviors within single castes of individual colonies would likely be absorbed in the error term in most studies. Perhaps this area deserves more careful study.

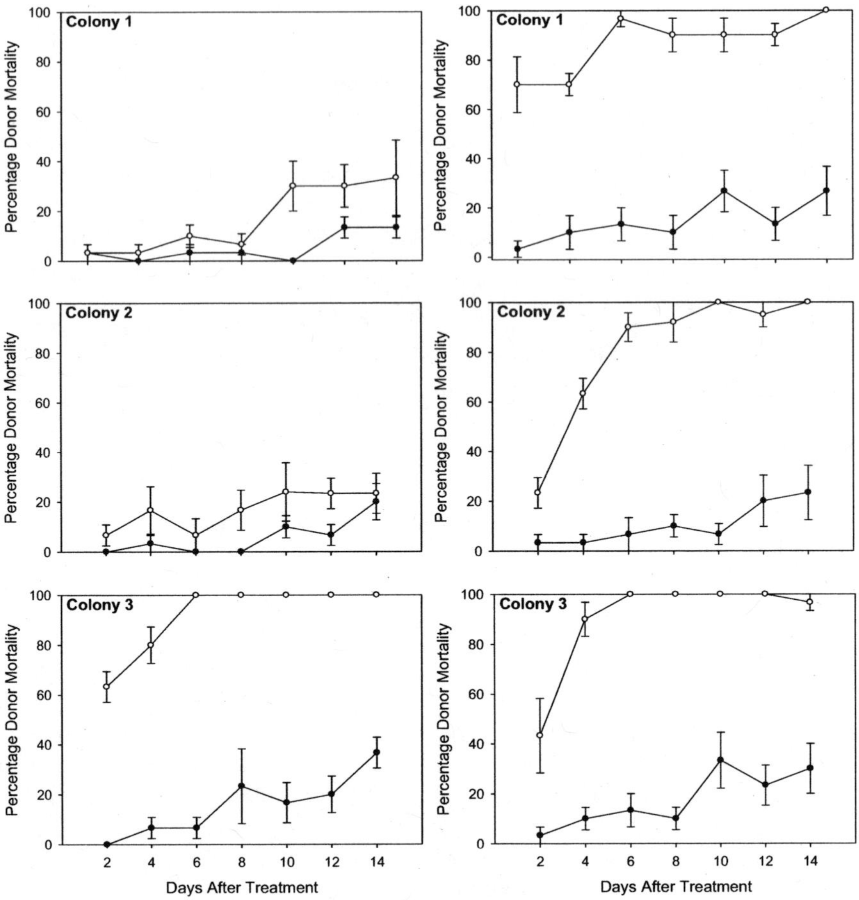

Figure 1. Donor mortality over time for each colony. Indoxacarb results are on the left and chlorantraniliprole results on the right in each column. Each data point is a mean ± SEM of 6 experimental units. For all graphs: ● are 0 ppm donors, and ○ are 100 ppm donors.

Failing to reject the consistency assumption as inaccurate, only the second possibility remains that intercolonial recipient mortality variability is not due to a vaguely defined behavioral variability among colonies. Other authors have already rejected the idea that this variability is correlated with body mass variability (*4, 6*). This presents a different problem: if the data given here suggest that behavioral variability among colonies is not the source of recipient mortality variation in transfer studies, then what is responsible? An explanation may be present in the data from the current study. Donor mortality also varies among colonies in these studies, but is consistent with the variation seen in the recipient mortality. In other words, the donors (who have been directly exposed

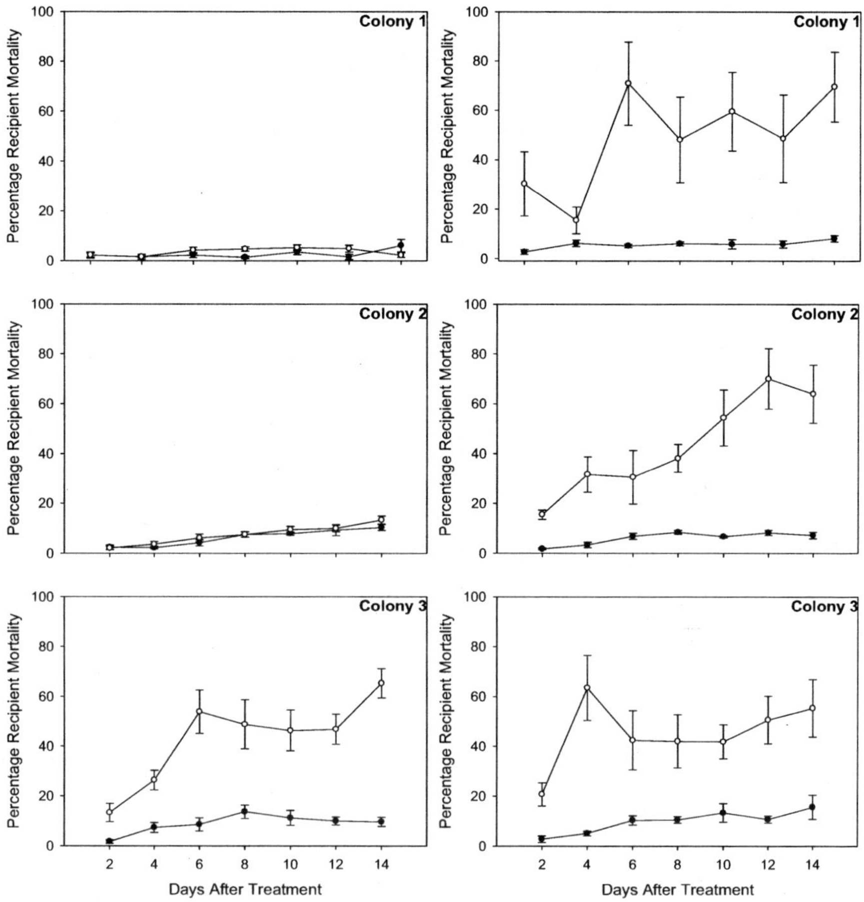

Figure 2. Recipient mortality over time for each colony. Indoxacarb results are on the left and chlorantraniliprole results on the right in each column. Each data point is a mean ± SEM of 6 experimental units. For all graphs: ● are 0 ppm donors, and ○ are 100 ppm donors.

to the termiticides) in colonies 1 and 2 are not effected as quickly as colony 3 donors, nor to the extent of colony 3 individuals (100% donor mortality at day 14 for colony 3 versus. < 40% donor mortality for colonies 1 and 2 at day 14). Admittedly, donor mortality for all three colonies was significantly influenced by both compounds over the entire course of the study. However, the data (Figure 1) indicates a very weak influence over controls for colony 2, and only slightly stronger for colony 1. In all, it appears that colonies 1 and 2 were less susceptible to indoxacarb, both when directly exposed (donors) and when exposed through transfer (recipients). The concept of variations in susceptibility among termite colonies has been investigated by Osbrink *et al.* (*29*) for *R. flavipes* and *C. formosanus*.

One of the qualifications when dealing with mortality as a measurement is the problem of dealing with unhealthy colonies, a question of colony vigor. It has been suggested that perhaps termite colony vigor is not binary (*i.e.*, healthy or not healthy), but is instead a spectrum ranging from very healthy colonies to very unhealthy colonies (*30*). It is possible that any variability observed in mortality among stressed individuals could possibly be the result of a slight stress acting in concert with low vigor to induce mortality. The opposite should also be true, where slight stresses may not induce mortality in very healthy colony members. Vigor-related influences in studies such as this are difficult (if not impossible) to distinguish unless the colony is in such poor health that high mortality occurs in the controls, otherwise mortality appears to result completely from the influence of treatment. However, the possibility of vigor differences bears mentioning whenever "colony effects" are noticed in termite studies. Recent studies have examined both means of determining vigor in laboratory termite colonies (*31*), as well as surveyed possible variables for this purpose (*32*). In the current study, control recipient mortality (Figure 2) did not indicate reduced vigor in the colonies.

Interestingly, while two colonies did not respond to indoxacarb in this study, no colonies were unresponsive to chlorantraniliprole. While both of these compounds are SC formulations, they belong to different classes and have different modes of action. Additionally, indoxacarb is a pro-insecticide, and is less toxic than its *N*-decarboxymethoxyllated metabolite (*33, 34*). Activity for indoxacarb is greatest when ingested by Lepidopteran larvae; and LD_{50}'s for oral *vs.* topical applications vary almost three-fold for certain larval Lepidoptera, although this does not hold for Coleoptera (*34*). If Saran and Rust (*10*) are correct about trophallaxis, it would be expected that only the parent compound of indoxacarb, rather than the toxic metabolite, is moved by recipients grooming donors. Indoxacarb's metabolite is active against insect Na^+ channels, disrupting action potentials (*34, 35*). In contrast the family that chlorantraniliprole belongs to, the anthranilic diamides, act against the ryanodine receptor (RyR) channels which control Ca^+ entry during muscle contraction events (*35*). Cordova *et al.* (*35*) examined 12 anthranilic diamides (of four classes) and compared them with indoxacarb. LD_{50}'s (oral) ranged from 0.4 to >500 ppm, compared with 0.6 ppm for indoxacarb in *Heliothis virescens* (Fabricius) larvae (*35*).

It is difficult to see an obvious reason for the apparent variability in susceptibility among colonies with indoxacarb that would not apply to chlorantraniliprole. It may be that the time necessary to convert indoxacarb from parent to metabolite differs among colonies, although recipient mortality for colonies 1 and 2 give no indication of increasing (beyond that of controls) even by day 14. Perhaps more detailed examinations of relative toxicity (oral and topical) of both compounds against subterranean termites is needed.

This chapter has investigated the source of the intercolonial variability in recipient mortality observed in toxicant transfer studies against subterranean termites in the laboratory. In summary, intercolonial differences in behavior are unlikely to be responsible for this variation in mortality. Instead it would appear that colonies vary in their physiological susceptibility to the compounds (as seen in directly exposed individuals), and that the absence of mortality is not necessarily correlated with either the presence or absence of transfer. Transfer may occur in colonies that are not susceptible to a particular toxicant, but it is not manifested by recipient mortality.

Acknowledgments

The author thanks Craig D. Bell and Shawn M. Cooper for expert technical assistance during these studies. The statistical assistance of Patrick D. Gerard (now of Clemson University) and Dennis E. Rowe (Mississippi State University) was greatly appreciated. The author also thanks two anonymous reviewers for their assistance in improving prior versions of this manuscript.

References

1. Kard, B.M. *Pest Control* **2001**, *69*, 30-33.
2. Tomalski, M.; Vargo, E.L. *Pest Control* **2004**, *72(5),* 51-53.
3. Ferster, B.; Scheffrahn, R.H.; Thoms, E.M.; Scherer, P.N. *J. Econ. Entomol.* **2001**, *94*, 215-222.
4. Thorne, B.L.; Breisch, N.L. *J. Econ. Entomol.* **2001**, *94*, 492-498.
5. Ibrahim, S.A.; Henderson, G.R.; Huixin, F. *J. Econ. Entomol.* **2003**, *96*, 461-467.
6. Shelton, T.G.; Grace, J.K. *J. Econ. Entomol.* **2003**, *96,* 456-460.
7. Hu, X.P.; Song, D.; Scherer, C.W. *Pest. Manage. Sci.* **2005**, *61*, 1209-1214.
8. Rust, M.K.; Saran, R.K. *J. Econ. Entomol.* **2006**, *99*, 864-872.
9. Shelton, T.G.; Mulrooney, J.E.; Wagner, T.L. *J. Econ. Entomol.* **2006**, *99*, 886-892.
10. Saran, R.K.; Rust, M.K. *J. Econ. Entomol.* **2007**, *100*, 495-508.
11. Shelton, T.G.; Bell, C.D.; Wagner, T.L. *Sociobiology* **2005**, *45*, 69-75.
12. Vargo, E.L.; Parman, V. *Pest Control* **2004**, *72(2)*, 36-38.
13. Osbrink, W.L.A.; Cornelius, M.L.; Lax, A.R. *J. Econ. Entomol.* **2005**, *98*, 2160-2168.
14. Sheets, J.J.; Karr, L.L.; Dripps, J.E. *J. Econ. Entomol.* **2000**, *93*, 871-877.
15. Myles, T.G. *Sociobiology* **1996**, *28*, 373-457.
16. Howse, P.E. *Ins. Soc.* **1968**, *15*, 45-50.
17. Maistrello, L.; Sbrenna, G. *Sociobiology* **1998**, *31*, 91-104.
18. Crosland, M.W.J.; Traniello, J.F.A. *Naturwissenschaften* **1997**, *84*, 208-211.
19. Crosland, M.W.J.; Lok, C.M.; Wong, T.C.; Shakarad, M.; Traniello, J.F.A. *An. Behav.* **1997**, *54*, 999-1012.
20. Lenz, M; Kard, B.; Mauldin, J.K.; Evans, T.A.; Etheridge, J.L.; Abbey, H.M. *International Research Group on Wood Protection (IRGWP)* **2000**, 1-8.
21. Hostettler, N.C.; Hall, D.W.; Scheffrahn, R.H. *Fla. Entomol.* **1995**, *78*, 119-129.

22. Su, N.-Y.; Ban, P.M.; Scheffrahn, R.H. *Sociobiology*, **1991**, *19*, 349-362.
23. SAS Institute *SAS user's guide: statistics*; SAS Institute, Inc.: Cary NC, 1985.
24. Plowright, R.C.; Plowright, C.M.S.; In: *Interindividual behavioral variability in social insects*, Jeanne, R.L., Ed., Westview Press, London, 1988, pp. 419-431.
25. Campora, C.E.; Grace, J.K. *J. Ins. Behav.*, **2004**, *17*, 777-791.
26. Su, N.-Y.; Haverty, M.I. *J. Ins. Behav.* **1991**, *4*, 115-128.
27. Thorne, B.L.; Haverty, M.I. *Sociobiology*, **1991**, *19*, 115-145.
28. Shelton, T.G.; Grace, J.K. *Sociobiology*, **1996**, *28*, 155-176.
29. Osbrink, W.L.A.; Lax, A.R.; Brenner, R.J. *J. Econ. Entomol.* **2001**, *94*, 1217-1228.
30. Lenz, M. *Bull. Entomol. Res.* **1985**, *75*, 13-21.
31. Arquette, T.J.; Forschler, B.T. *J. Econ. Entomol.* **2006**, *99*, 873-878.
32. Arquette, T.J.; Champagne, D.E.; Brown, M.R.; Forschler, B.T. *J. Ins. Physiol.* **2006**, *52*, 51-66.
33. Wing, K.D.; Andalaro, J.T.; McCann, S.F.; Salgado, V.L. In: *Comprehensive molecular insect science*, Vol. VI, Gilbert, L.I.; Iatrou, K.; Gill, S.S., Eds., Pergamon Press, Oxford, **2004**, pp. 31-53.
34. Wing, K.D.; Sacher, M.; Kagaya, Y.; Tsurubuchi, Y.; Mulderig, L.; Connair, M.; Schnee, M. *Crop Protection* **2000**, 19, 537-545.
35. Cordova, D.; Benner, E.A.; Sacher, M.D.; Rauh, J.J.; Sopa, J.S.; Lahm, G.P.; Selby, T.P.; Stevenson, T.M.; Flexner, L.; Gutteridge, S.; Rhoades, D.F.; Wu, L.; Smith, R.M.; Tao, Y. *Pesticide Biochem. & Physiol.* **2006**, 84, 196-214.

Chapter 7

Biological Activities of a Bait Toxicant for Population Management of Subterranean Termites

Nan-Yao Su[1] and Mike Lees[2]

[1]Department of Entomology and Nematology, Ft. Lauderdale Research & Education Center, University of Florida, Ft. Lauderdale, FL
[2]Dow AgroSciences, Granite Bay, CA

Termites continue to be a source of concern for property owners due to the potential for damage. The use of baiting for control of damaging subterranean termites has been researched for many years and utilized commercially for more than a decade. Several recent factors have driven interest in the development of baiting systems. The understanding of population dynamics of subterranean termites is important in developing and implementing successful baiting programs. Numerous studies have shown that proper introduction of toxicants in bait matrices into termite colonies can eliminate entire colonies. This is different than the exclusion of termites by using chemically treated soil barriers, which can lead to future infestations or infestation to adjacent properties. When evaluating bait toxicants, several factors should be considered. The toxicant should be slow-acting so that it may be transferred throughout the colony before the onset of toxic symptoms. A toxicant should also be nonrepellent and palatable to termites when combined with a matrix of a desired food source. Finally, the lethal time of an ideal toxicant should be dose-independent so that high doses can be accumulated in a colony population without a direct correlation to mortality. The combination of these factors allows an active ingredient to successfully spread throughout a colony and affect individual mortality, colony decline and eventual colony elimination.

The predominant business model for the termite control industry is to protect the structure under contract from subterranean termites. Soil termiticide barriers have been used for this purpose for more than half-century by excluding soil-borne termites from the structure. An industry survey conducted in 2002, for example, indicated that soil termiticide applications accounted for a 77% market share of the subterranean termite control business in the United States (*1*), and this figure has increased since then. A subterranean termite colony may contain 100,000 to more than 1,000,000 individuals with a foraging territory extending up to 100 m (*2, 3, 4*). Despite the application of the large quantity of insecticide (5-10 kg per house), soil termiticide barriers usually do not affect the vast populations of subterranean termites around the structure (*2*). The surviving colonies, or portions of colonies, not affected by termiticide barriers may move on to infest structures in the vicinity and produce alates that further infest nearby areas. Despite this potential for future infestations, the main method of termite control remains the barrier treatment. This reliance on soil termiticide barriers is one contributing factor for the continuing expansion of the Formosan subterranean termite, *Coptotermes formosanus* Shiraki, in the United States (*5*).

Another factor influencing termite control methods is the increased scrutiny of pesticide use by the public and calls for implementation of more environmentally responsible methods of subterranean termite control, such as the increasing popularity of Integrated Pest Management (IPM) programs (*6, 7, 8*). In addition to elimination of termite colonies as opposed to solely exclusion, baiting also addresses environmental concerns by relying on much lower volumes of toxicant as well as application of toxicants on an as-needed basis as opposed to prophylactic treatments to large areas.

To properly assess and implement a termite baiting strategy, several criteria should be evaluated. An understanding of population biology of subterranean termites is needed in order to construct a system that will impact the entire colony. Key considerations when evaluating a potential active ingredient as a bait toxicant are repellency of the toxicant and palatability in combination with a matrix of feeding substances, length of time and dosage required to cause mortality. These parameters are reviewed with respect to several categories of bait toxicants.

Population management

One requirement for an IPM program to control subterranean termites, as proposed by Su and Scheffrahn (*6*), was to reduce termite damage potential by managing their populations. Instead of merely excluding subterranean termites from individual houses, the focus of an IPM program is the area-wide management of subterranean termite populations. Population management of subterranean termites targets the colony instead of termite individuals. A colony is defined as "a group of termites sharing interconnected foraging sites" (*6*). Studies show that partially suppressed colonies may recover over time and cause additional damage whereas elimination of target colonies usually creates a zone of termite-free soil that lasts for months or years (*9*). Thus, successful management of subterranean termite populations requires the application of a control measure that is capable of eliminating the target colonies.

There have been several area-wide pilot projects for population management of subterranean termites in recent years (*10, 11, 12, 13, 14, 15, 16*), and most depended heavily on colony-eliminating baits that include the chitin synthesis inhibitors (CSIs) such as hexaflumuron or noviflumuron. Approximately 70% of treatments for the Operation Full Stop (a national program for population management of the invasive Formosan subterranean termite by USDA-ARS), for example, employed the Sentricon® *Termite Colony Elimination System* (Dow AgroSciences LLC, Indianapolis, IN) that contains hexaflumuron or noviflumuron as the active ingredient (AI) (*17*). With the Sentricon® System, Ross (*15*) reported the eradication of an isolated infestation of the invasive *C. acinaciformis* (Frogatt) from a rural town of the North Island of New Zealand. Significant reduction of *Reticulitermes* populations was recorded from a large community of 132 buildings over an area of 90 acres following the extensive application of the Sentricon® System (*13*). The Sentricon® System was also used successfully in eliminating the populations of the invasive *R. flavipes* (Kollar) from several town blocks of a low-income community in Santiago, Chile (*16*). The successful outcomes of these pilot projects are indicative of the importance of the ability of the control measure to eliminate the subterranean termite colonies.

Repellency, deterrence, and lethality

Early research into baiting for control of termites led to a refined understanding of toxicant attributes related to efficacy. Esenther and Gray (*18*) suggested that wooden blocks impregnated with slow-acting toxicants such as dechlorane (mirex) might be used to eliminate colonies of subterranean termites. Subsequent studies with mirex-baits indicated that a continuous placement of toxic baits may suppress foraging activities of *R. flavipes* (*19, 20, 21, 22*), but the effects of mirex-baits on colony populations were not assessed. The importance of slow activity is to increase the chance of transfer of the AI between colony mates. Bait toxicants take advantage of trophallaxis by termites. Faster acting AIs may cause mortality before sufficient transfer occurs and prevent total colony elimination. In addition to the slow-acting characteristic, Su et al. (*23*) considered that an AI had to be nonrepellent in order to eliminate colonies of subterranean termites. Varying degrees of repellency may prevent introduction of sufficient quantities of toxicant into the colony.

AI repellency and feeding deterrence (if the AI is to be incorporated into a feeding substance as baits) can be determined in a laboratory test in which termites are provided with a choice to avoid treatments (*23, 24*). AI lethality, repellency and deterrence are inter-related and concentration-dependent. Before carrying out an expensive field trial, a laboratory choice test has to be conducted to determine if the AI can yield significant mortality at a concentration range that does not cause repellency or deterrence (*24, 25, 26*). Assuming that a hypothetical AI causes significant mortality at concentrations of $> \alpha$ (mortality threshold concentration) in a laboratory choice test, whereas substantial repellency or feeding deterrence was observed with concentrations of $> \beta$ (deterrence threshold concentration). If $\alpha < \beta$, then the AI may have potential for a successful field trial, and it may be more useful when the difference between α and β is larger, or $\alpha << \beta$. On the other hand, if $\alpha > \beta$, then it has to be assumed that termites may avoid the AI treatments before sustaining substantial mortality.

In a laboratory choice test by Su and Scheffrahn (*26*), only feeding blocks impregnated with >1,000 ppm acetone solution of diflubenzuron yielded significant mortality for *C. formosanus* (thus the mortality threshold concentration α = 1,000 ppm), while diflubenzuron concentrations as low as 2 ppm deterred feeding of this species, i.e. the deterrence threshold concentration β = 2 ppm (Table 1). With *R. flavipes*, α for diflubenzuron was 7.8 ppm and β was 31.3 ppm. For hexaflumuron, α and β for *C. formosanus* were 15.6 and 125 ppm, respectively, and α and β for *R. flavipes* were 2 and 62.5 ppm, respectively (Table 1). Because $\alpha > \beta$ for diflubenzuron against *C. formosanus,* Su and Scheffrahn (*25*) concluded that diflubenzuron is not likely to be an effective bait toxicant against this termite species. Conversely, since $\alpha < \beta$ for *R. flavipes* for diflubenzuron, it was expected to be an effective toxicant. Diflubenzuron has been used in commercial bait products against subterranean termites in the last decade, but thus far no field data are available to demonstrate its ability to eliminate colonies of *C. formosanus.*

Table 1. Minimum AI concentrations that caused significant mortality (α: mortality threshold concentration) or feeding deterrence (β: deterrence threshold concentration) in a laboratory choice test in which two AIs, diflubenzuron and hexaflumuron, were impregnated in feeding blocks for *C. formosanus* and *R. flavipes*. α and β are AI concentrations (ppm) in acetone solution used for impregnation, wt (AI) / cc acetone

AI	*C. formosanus*		*R. flavipes*	
	α	β	α	β
Diflubenzuron	> 1,000	2	7.8	31.3
Hexaflumuron	15.6	125	2	62.5

Based on data of Su and Scheffrahn (*26*)

In another laboratory choice test using sawdust baits impregnated with lufenuron or hexaflumuron, Su and Scheffrahn (*27*) reported that α for lufenuron against *C. formosanus* was > 8,000 ppm (AI wt / dry wt bait), and β = 2,000 ppm, and α and β for *R. flavipes* were 800 and 100 ppm, respectively (Table 2). Because $\alpha > \beta$ for lufenuron against both termite species, Su and Scheffrahn (*27*) concluded that lufenuron was not a good candidate as a bait toxicant. It should be noted that mortality data (α) were taken at the end of these 9-wk choice tests, whereas the feeding deterrence data (β) were taken at 3 wk (*26, 27*). In both laboratory studies, α was consistently less than β for hexaflumuron against both *C. formosanus* and *R. flavipes*. The importance of $\alpha < \beta$ was confirmed by numerous field studies that demonstrated eliminations of subterranean termite colonies by hexaflumuron (*28*). Due to the substantial cost, it may not be logical to field test an AI with $\alpha > \beta$ and it is even riskier to commercialize a bait product using such an AI.

Table 2. Minimum AI concentrations that caused significant mortality (α: mortality threshold concentration) or feeding deterrence (β: deterrence threshold concentration) in a laboratory choice test in which two AIs, lufenuron and hexaflumuron, were homogenized in sawdust baits for *C. formosanus* and *R. flavipes*. α and β are AI concentrations (ppm) in dry bait, wt (AI) / wt dry bait

AI	*C. formosanus*		*R. flavipes*	
	α	β	α	β
Lufenuron	> 8,000	2,000	800	100
Hexaflumuron	125	20,000	31.3	8,000

Based on data of Su and Scheffrahn (*27*)

Lethal time and AI dose

Although delayed toxicity ("slow-acting") is recognized as one important characteristic of an AI for population control of social insects such as ants and termites, there is no consensus for its definition, or how to measure the lethal time of a potential AI. The delayed toxicity of a bait toxicant for control of the red imported fire ant, *Solenopsis invicta* Buren, for example, was defined as "<15% mortality at 1 d and >89% at 14 d" (*29*). This definition was adopted by many for screening of bait toxicants for *S. invicta* (*30, 31, 32, 33*). Using the Weibull function, Haverty and Dell (*34*) estimated the time required to achieve 90% mortality of the pine cone beetle, *Conophthorus ponderosae* Hopkins.

Su et al. (*35*) quantified lethal time of potential bait AIs against *C. formosanus* by measuring the time required for an AI to fully express its effects. When exposed to a high concentration of fast-acting AIs (chlordane or chlorpyrifos) under a no-choice test (thus the AI concentration [ppm] was more or less proportional to the dose [AI wt / termite]), all termites were killed within hours. The reduced doses of "fast-acting" AIs resulted in lower final mortalities but the time required to reach respective mortality remained more or less the same (lethal time = T_h) for all doses (Fig. 1A). As with the "fast-acting" AIs, the high dose of supposedly "slow-acting" AIs (amidinohydrazone and avermectin B1) also killed all termites within hours, but when the doses were reduced, the final mortalities were reached at protracted time frames of T_{mh} and T_{lh} for medium-high dose and medium-low dose, respectively (Fig. 1B). The distinctive difference between the fast- and slow-acting AIs was the delayed mortalities produced by the reduced doses of the latter. Early on, it was assumed that a proper AI concentration may be chosen to produce reduced doses when consumed by termites which would ultimately result in the protracted time frames of T_{mh} and T_{lh} (*24, 25*), but this assumption turned out to be untrue for some AIs.

Dose-dependent lethal time for metabolic inhibitors and nonrepellent termiticides

The concept of dose dependency is important in assessing bait toxicants. A successful toxicant can be distributed throughout a colony affecting enough workers to cause colony collapse and resultant elimination. Through this process, varying concentrations of toxicant will be present in individuals. The effect of the toxicant related to concentration in an individual may affect the transfer of that AI.

Based on the studies of Su et al. (*23, 25*), several metabolic inhibitors were identified to be slow-acting and nonrepellent, including hydramethylnon (*23*), A-9248 (diiodomethyl para-tolyl sulfone) (*24*), and sulfluramid (*36*). None of these metabolic inhibitors, however, successfully eliminated field colonies of subterranean termites (*37, 38, 39, 40*). These field results led Su et al. (*38*) to conclude that the inability of metabolic inhibitors to eliminate populations or foraging activity of target colonies was due to their dose-dependent lethal time, i.e., time required to death after exposure to lethal dose.

By determining the deterrence threshold concentration (β) in a laboratory choice test, and the concentrations that may result in delayed mortalities (T_{mh} and T_{lh}, Fig. 1B), a toxicant concentration in baits may be chosen for a field trial so that baits are accepted by termites, but the total amount of bait ingested by termites cannot be manipulated. Following bait placement in a colony, those ingesting a large quantity of bait may contain high doses, while other termites may acquire medium to low does, sub-lethal dose, or no ingestion at all. Because lethal time for metabolic inhibitors is dose-dependent (Fig. 1B), termites ingesting a high lethal dose may be killed relatively quickly (lethal time = T_h, Fig. 1B), thus negating the slow-acting characteristic required for effective bait transfer and assimilation (*6*). Only those ingesting medium doses may be killed slowly after walking away from treatment zones (lethal time = T_{mh} or T_{lh}, Fig. 1B). Results of field trials suggest that termites ingesting sub-lethal doses of these metabolic inhibitors may "learn" to avoid feeding on treated baits (*38*).

Recent results with nonrepellent termiticides also indicate the importance of understanding the relationship between dose and lethal time. Nonrepellent liquid termiticides applied as a barrier treatment around the perimeter of targeted structures have been the popular choices for termite control industry in recent years. The 2002 industry survey (*1*), for example, showed that ≈60% of the termiticides used in the United States were one of the nonrepellent termiticides such as fipronil (Termidor®, BASF Corp. Research Triangle Park, NC), imidacloprid (Premise®, Bayer Environmental Service, Montvale, NJ), or chlorfenapyr (Phantom®, BASF Corp.). Due to their nonrepellency and apparent delayed action, it has been suggested that these termiticides may impact the subterranean termite populations through a horizontal transfer (*41, 42, 43, 44, 45*). Ibrahim et al. (*46*) and Hu (*45*) indicated that movement of exposed termites may spread the nonrepellent toxicants to nestmates through trophallaxis and social grooming, but Shelton and Grace (*47*) reported that high concentration of fipronil (> 10 ppm) was needed for a successful transfer of lethal dose to recipients, and at such dose, the donors may be killed too quickly for a substantial toxicant transfer to occur within the population. Using an extended foraging arena to simulate the distance factor of a field colony of subterranean termites, Su (*48*) reported that the horizontal transfer of lethal effects of fipronil was ≤ 5 m. Saran and Rust (*49*), who studied the horizontal transfer of ^{14}C-radiolabeled fipronil in laboratory groups of *R. hesperus* Banks, concluded that exposed termites were too severely impaired to be mobile, and such transfer was not a major factor contributing to field efficacy of fipronil.

These laboratory results appear to corroborate field studies with nonrepellent termiticides. When field efficacy of fipronil was measured by using monitoring stations, Potter and Hillery (*43*) reported that termite activity 0.3 – 4 m from the treatments were eliminated (at six of eight sites), but those of > 5 m away from treatments remained active with termites. Osbrink et al. (*50*) reported that monitoring stations 1 – 3 m away from soil treated with imidacloprid were unaffected by the treatment. Another field study to evaluate potential control measures against *R. flavipes* in Chile also showed no significant change in termite foraging activity in sites treated with fipronil (*40*). Results of these studies indicated that lethal times of nonrepellent termiticides are similar to those of other metabolic inhibitor bait AIs such as hydramethylnon, A-9248, or sulfluramid, and their failures to eliminate a colony are due to their dose-dependent lethal times (Fig. 1B).

Dose-independent lethal time for chitin synthesis inhibitors (CSIs)

Contrary to metabolic inhibitor baits or nonrepellent termiticides, numerous studies have demonstrated the elimination of all detectable subterranean termite activity by CSI (hexaflumuron or noviflumuron) baits at a variety of locations with a variety of different termite species. Between 1994 and 2001, for exmaples, 33 field studies with hexaflumuron demonstrated elimination of 152 of 159 baited colonies or populations (96% success rate) including 13 termite species worldwide (*28*). More field studies, including area-wide IPM pilot projects with hexaflumuron or noviflumuron baits, have reported similar positive results with these two CSIs since 2001 (*10, 11, 12, 13, 14, 15, 16*).

The main difference between metabolic inhibitors and CSIs is their lethal time, i.e. time required to death after exposure to lethal doses (*6*). CSIs mainly affect termites through disruption of their molting process, so regardless of AI doses, termites are not affected until they molt. Thus even after ingesting a high dose of a CSI, mortality is not fully expressed until much later (lethal time = T_h, Fig. 1C), presumably when the molting process was inhibited. As long as a lethal dose is ingested, the high, medium-high or medium-low doses would have produced a more or less similar lethal time (Fig. 1C), which is dependent on the timing of ecdysis for individual termites. The most important aspect of a CSI is that even at the high dose, the lethal time does not resemble those of the fast or slow-acting metabolic inhibitors, T_h (Fig. 1). Due to the overt toxicity of an insect growth regulator, such as CSIs, a higher lethal dose might result in slightly faster kill (Fig. 1C). Even with such differences among the doses, however, lethal times produced by CSIs typically range from weeks to months instead of days for slow-acting metabolic inhibitors, and hours for fast-acting AIs. Such a protracted lethal time that is relatively independent of AI doses in termites during the baiting period is probably the vital factor contributing to the success of eliminating colonies of subterranean termites as reported by numerous field studies. In addition to being "slow-acting" and "nonrepellent," lethal time of an AI has to be dose-independent if it is to eliminate the vast colony of subterranean termites.

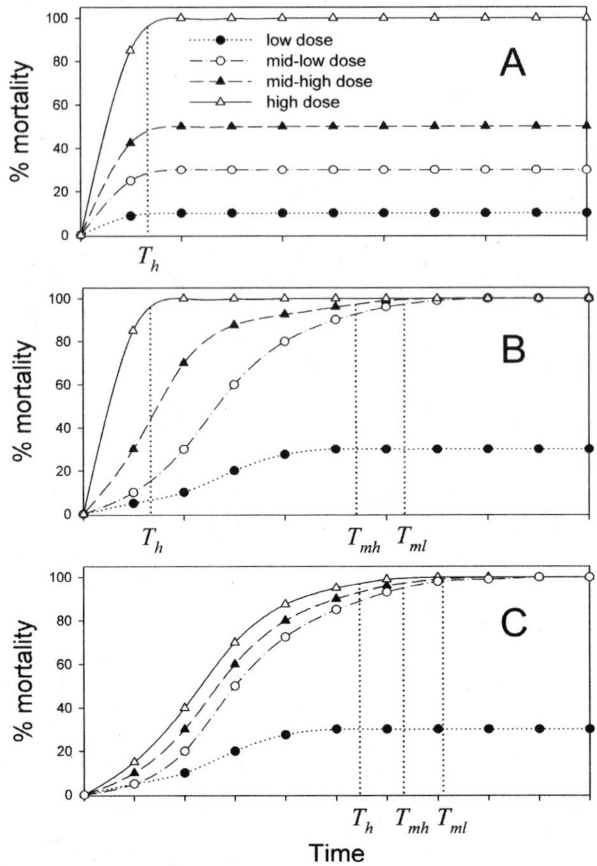

Figure 1. The relationship between lethal time and dose of fast-acting AIs (A), slow-acting metabolic inhibitors (B), and chitin synthesis inhibitors (CSIs) such as hexaflumuron or noviflumuron (C). Lethal time is defined as the time for an AI to fully express its effects against test insects. At the high doses, both fast-acting AIs and slow-acting metabolic inhibitors cause substantial mortalities quickly (T_h, A and B). The reduced doses of a fast-acting AI produce lower final mortalities, but the lethal time remains the same for all doses at T_h (A). When the doses are reduced for slow-acting metabolic inhibitors, the final mortalities are reached at protracted time frames of T_{mh} and T_{lh} for medium-high dose and medium-low dose, respectively (B). CSIs mainly affect termite molting, and the timing of ecdysis is independent of doses. Thus as long as a lethal dose is ingested, the high, medium-high or medium low doses would have produced more or less the similar lethal time (C).

Acknowledgments

This research was supported in part by a grant from USDA-ARS under the grant agreement No. 58-6435-3-0075.

References

1. Anonymous. *Pest Control* **2002**, 70, S1-S23.
2. Su, N.-Y.; Scheffrahn, R. H. *Sociobiology* **1988**, 14, 353-359.
3. Grace, J. K.; Abdallay, A.; Farr, K. R. *Can. Entomol.* **1989**, 121, 551-556.
4. Su, N.-Y.; Ban, P. M.; Scheffrahn, R. H. *Environ. Entomol.* **1993**, 22, 1113-1117.
5. Su, N.-Y. *Sociobiology* 2003, 41, 7-16.
6. Su, N.-Y.; Scheffrahn, R. H. *Integrated Pest Management Reviews* **1998**, 3, 1-13.
7. Forschler, B. T.; Jenkins, T. M. *Urban Ecosystems* **2000**, 4, 231-251.
8. Williams, G. M. *J. Econ. Entomol.* **2005**, 98, 1275-1283.
9. Su, N.-Y.; Scheffrahn, R. H. *Sociobiology* **1996**, 27, 253-275.
10. Ring, D. R.; Morgan, A. L.; Woodson, W. D. *Sociobiology* **2001**, 37, 293-300.
11. Lax, A. R.; Osbrink, W. L. A. *Pest Manag. Sci.* **2003**, 59, 788-800.
12. Su, N.-Y.; Ban, P. M.; Scheffrahn, R. H. *J. Econ. Entomol.* **2004**, 97, 2029-2034.
13. Getty, G. M.; Solek, C. W.; Sbragia, R. J.; Haverty, M. I.; Lewis, V. R. In *Proc. 5th Int'l. Conf. Urban Pests*; Lee, C.-Y.; Robinson, W. H., Eds., P&Y Design Network, Penang, Malaysia, 2005; pp. 165-169.
14. Guillot, F. S.; Ring, D. R.; Lax, A. R.; Boykin, D. In *Proc. 5th Int'l. Conf. Urban Pests*; Lee, C.-Y.; Robinson, W. H., Eds., P&Y Design Network, Penang, Malaysia, 2005; pp. 171-178.
15. Ross, M. G. In *Proc. 5th Int'l. Conf. Urban Pests*; Lee, C.-Y.; Robinson, W. H., Eds., P&Y Design Network, Penang, Malaysia, 2005; pp. 233-238.
16. Smith, J.; Su, N.-Y.; Escoba, R. *Am. Entomol.* **2006**, 52, 253-260.
17. Guillot, F. S. USDA-ARS, *unpublished.*
18. Esenther, G. R.; Gray, D. E. *Can. Entomol.* **1968**, 100, 827-834.
19. Beard, R. L. *Connecticut Agr. Exp. Sta. Bull.* **1974**, 748.
20. Esenther, G. R.; Beal, R. H. *J. Econ. Entomol.* **1974**, 67, 85-88.
21. Esenther, G. R.; Beal, R. H. *J. Econ. Entomol.* **1978**, 71, 604-607.
22. Ostaff, D.; Gray, D. E. *Can. Entomol.* **1975**, 107, 1321-1325.
23. Su, N.-Y.; Tamashiro, M.; Yates, J. R.; Haverty, M. I. *J. Econ. Entomol.* 1982, 75, 188-193.
24. Su, N.-Y.; Scheffrahn, R. H. *J. Econ. Entomol.* **1988**, 81, 850-854.
25. Su, N.-Y.; Scheffrahn, R. H. *J. Econ. Entomol.* **1991**, 84, 170-175.
26. Su, N.-Y.; Scheffrahn, R. H. *J. Econ. Entomol.* **1993**, 86, 1453-1457.
27. Su, N.-Y.; Scheffrahn, R. H. *J. Econ. Entomol.* **1996**, 89, 1156-1160.
28. Su, N.-Y. *Sociobiology* **2003**, 41, 177-192.
29. Stringer, C. E., Jr.; Lofgren, C. S.; Bartlett, F. J. *J. Econ. Entomol.* **1964,** 57, 941-945.

30. Williams, D. F.; Lofgren, C. S.; Banks, W. A.; Stringer, C. E.; Plumley, J. K. *J. Econ. Entomol.* **1980**, 73, 798-802.
31. Williams, D. F.; Lofgren, C. S. 1981. *Fla. Entomol.* **1981**, 64, 472-477.
32. Williams, D. F. *Fla. Entomol.* **1983**, 66, 163-172.
33. Vander Meer, R. K.; Lofgren; C. S.; Williams, D. F. *J. Econ. Entomol.* **1986**, 79, 1190-1197.
34. Haverty, M. I.; Dell, T. R. *Pestic. Sci.* **1984,** 15, 369-374.
35. Su, N.-Y.; Tamashiro, M.; J. R.; Haverty, M. I. *J. Econ. Entomol.* **1987**, 80,1-4.
36. Su, N.-Y.; Scheffrahn, R. H. *Florida Entomol.* **1988**, 71, 73-78.
37. Su, N.-Y.; Ban, P. M.; Scheffrahn, R. H. *J. Econ. Entomol.* **1991**, 84, 1524-1531.
38. Su, N.-Y.; Scheffrahn, R. H.; Ban, P. M. *J. Econ. Entomol.* **1995**, 88, 1343-1348.
39. Pawson B. M.; Gold, R. E. *Sociobiology* **1996**, 28, 485-510.
40. Ripa, R.; Luppicini, P.; Su, N.-Y.; Rust, M. K. *J. Econ. Entomol.* **2007**, 100, 1391-1399.
41. Kard, B. *Pest Control* **2001**, 69, 30-33.
42. Thorne, B. L.; Breisch, N. L. 2001. *J. Econ. Entomol.* **2001**, 94, 492-498.
43. Potter, M. H.; Hillery, A. E. *Sociobiology* **2002**, 39, 373-405.
44. Wagner, T. *Sociobiology* **2003**, 41, 131-141.
45. Hu, X. P. *J. Econ. Entomol.* **2005**, 98, 509-517.
46. Ibrahim, S. A.; Henderson, G.; Fei, H. *J. Econ. Entomol.* **2003**, 96, 461-467.
47. Shelton, T. G.; Grace, J. K. *J. Econ. Entomol.* **2003**, 96, 456-460.
48. Su, N.-Y. *J. Econ. Entomol.* **2005**, 98, 2143-2152.
49. Saran, R. K.; Rust, M. K. *J. Econ. Entomol.* **2007**, 100, 495-508.
50. Osbrink, W. L. A.; Cornelius, M. L.; Lax. A. *J. Econ. Entomol.* **2005**, 98, 2160-2168.

Chapter 8

Depth of initial penetration of two aqueous termiticide formulations as a function of soil type and soil moisture

Chris J. Peterson

USDA Forest Service, Insects, Diseases and Invasive Plants Research Unit, 201 Lincoln Green, Starkville, MS 39759

The initial penetration of two termiticide formulations, Premise 75 (imidacloprid) and Termidor SC (fipronil), were tested in four soils at three moisture levels (5, 10 and 15% by weight) in a laboratory study. Within each soil type and moisture combination, the highest concentrations of active ingredient were found in the top 1 cm of soil and decreased with increasing depth. As soil moisture increased, active ingredient concentration in the top 1 cm decreased while active ingredient concentration in lower depths, especially 2 to 5 cm, increased. For each compound, the effect of soil type on active ingredient penetration depended on the soil moisture and soil depth, with few effects at low moisture and greater depth. Soil type had little overall effect on the penetration of either compound, however, as both compounds were contained in the top 5 cm in each situation. Both compounds were the most toxic to termites in soils with low organic matter.

Chemical soil treatment for the prevention of termite infestation in structures has been practiced since at least the late 1920s (*1*), with previous recommendations relying solely on good building practices (such as minimizing soil-wood contact) and impregnated timber (*2*). The first tests of soil chemical application were initiated in 1928 in California using termite-infested utility poles (*3*). It is interesting that chemical soil treatment, now a multi-billion dollar industry in the United States, was originally thought of as being useful only on a temporary basis (*4*) and should not substitute for good building practices. Good building practices are still recommended in addition to chemical application (*5*) and are incorporated into most building codes.

Prior to the end of the Second World War, most houses were of the wall and pier, or "conventional" foundation type. Soil treatment in this type of construction consists of trench applications, where soil is removed around a foundation wall or support piers in a trench about six inches (15 cm) wide, and then the soil is treated as it is being replaced. This method is still used for what are now referred to as "perimeter" treatments.

Following the Second World War, houses constructed on a concrete slab in direct contact with the ground began to gain in popularity (*6*) and continue to do so. According to the United States Census Bureau, 72% of all houses built in the southern United States in 2006 had slab foundations, compared with 46% in 1971, the first year for which records of this type were available. Some thought slab construction was an end to termite problems, because termites would not be able to penetrate several inches of concrete. It was believed that a perimeter treatment around the slab would prevent attacks from the edge. However, it was soon found that termites could, and did, enter structures from below through plumbing and electrical service penetrations, expansion joints and cracks (*7*). It was therefore recommended that an overall termiticide application to the soil before the slab is poured would prevent termite access through these areas.

The United States Department of Agriculture - Forest Service (USFS) was among the first to test the efficacy of this application method. Based on tests initiated in 1946, an application rate of 1 pint of insecticide formulation per square foot (4.75 liter per square meter, or 1.25 gallons per 10 square feet, more than the current label rate) was proposed in 1954 (*7*). This was adjusted to 1 gallon per 10 square feet (4 liter per 1 square meter) in 1956, for the reason than it was simpler for the applicator to calibrate spraying equipment in gallons-per-minute and use simple math to determine how much solution was needed (or for how long to run the sprayer) once the square footage was known (*8*). For example, treating 1000 square feet would require 100 gallons and take 20 minutes at five gallons per minute. The Federal Housing Administration adopted this rate as a guideline in 1958 (*9*), and it is now considered the standard industry practice.

The integrity of the chemical barrier is important to the prevention of termite infestations. In slab-type construction, shortly after a termiticide is applied, a vapor barrier is placed over the soil, reinforcing bars or mesh is laid, and concrete is poured over the vapor barrier. These processes may take place over the course of several hours to more than one day, and all of these activities raise the potential of disturbance to the chemical barrier. If the soil disturbance is great, the integrity of the chemical barrier may be compromised. A "perfect" termiticide formulation should penetrate deeply enough to provide a barrier resistant to minor disturbance but not penetrate so deeply that the compound is diluted by soil to below the level of effectiveness.

The initial soil penetration of termiticide solutions has not been examined since around 1970, when USFS personnel studied the depth of initial penetration of organochlorine termiticides (*10 – 13*). These studies determined that most of the applied insecticide remained in the top 0.75 inch (2 cm) of the soil. The active ingredients used, chlordane, aldrin, dieldrin and heptachlor, are nearly insoluble in water and practically immobile in the environment, especially under the conditions found in termite control (i.e. beneath a concrete slab) where they are protected from the elements (*14*). In some cases, organochlorine insecticides were diluted in fuel oil or kerosene (for an example, see *15*), a practice no longer used. Even when diluted in water, the concentrated forms of these products contained petroleum distillates or hydrocarbons (for examples, see *16* and *17*).

Most termiticidal active ingredients introduced since about 1970 have been more water-soluble than earlier compounds, for example permethrin (<1 mg/L), chlorpyrifos (2 mg/L), fipronil (2 mg/L) and imidacloprid (510 mg/L) (*18*). Water-soluble compounds have a greater potential than insoluble compounds to move through the soil with the application solution. This may aid in the spread of the active ingredient, resulting in a more uniform distribution in the soil due to lateral and vertical movement. Hydrophobic compounds diluted in a petroleum carrier should penetrate the soil differently than more hydrophilic compounds diluted in water. Systematic evaluations of soil penetration by aqueous solutions of newer active ingredients have not been made.

This study examines the initial depth of penetration of two aqueous termiticide formulations, Premise and Termidor in four different soils and at three soil moisture levels.

Materials and Methods

Soils

Four soil types, designated U, D, H and P were collected, reflecting different contents of clay, silt, sand, organic matter, pH, cation exchange capacity (CEC) and field capacity (Table 1). U soil was loamy sand collected from the USFS Termiticide Testing Program site in Union County, SC. D soil was silt loam collected in the John Starr Memorial Forest near Dorman Lake in Oktibbeha County, MS. H soil, a sandy loam, was collected from the USFS Termiticide Testing Program site in the Harrison Experimental Forest in Harrison County, MS. P soil was sandy loam collected from Parker Sand and Gravel Co., Lowndes County, MS and is of a type approved by local building authorities for use as construction fill. All soils were air-dried, clumps were broken apart with a hammer and each soil was sieved to remove stones and roots. The soil texture analysis, pH, organic matter and cation exchange capacity was determined by the Mississippi State University Extension Service. To approximate the water holding capacity, 50-g portions of each soil (oven-dried at 100 °C overnight) were placed in Buchner funnels fitted with filter paper to prevent loss of soil. Distilled water, enough to thoroughly wet each soil, was added and a 34.5 kPa (5 psi) vacuum was applied until water was no longer observed dripping from the funnel. The soils were re-weighed and the water content was calculated (*19*).

Table 1. Properties of soils used in this study

Soil Type	Texture	Silt (%)	Sand (%)	Clay (%)	pH	%OM[a]	CEC[b]	Field Capacity (%)
U	Loamy Sand	19.75	77.75	2.50	5.2	1.41	4.10	16.6
D	Silt Loam	50.00	42.50	7.50	5.3	2.43	15.20	35.9
P	Sandy Loam	40.00	55.00	5.00	5.1	0.52	6.00	21.2
H	Sandy Loam	27.75	69.75	2.50	5.0	2.17	4.50	17.6

[a] Percentage organic matter
[b] Cation exchange capacity

To hydrate each soil for the test, the mass (± 0.1 kg) of each soil in a 19 liter (5 gal) bucket was found, and the amount of water required to constitute three moisture levels (5, 10 and 15% by weight) was calculated. Water was added to each 19 liter soil portion in a cement mixer by using a carbon dioxide sprayer during tumbling in a cement mixer for > 5 min. The soil for each 19-liter portion was then added to six plastic buckets (18 × 14 ID) to a depth of 15 cm.

Soil treatment, extraction and analysis

Two commonly used termiticides, Termidor and Premise, were mixed at the labeled rate for sub-slab treatment (0.06% and 0.05%, respectively). The application of the termiticide solutions was conducted within two hours of soil hydration. The termiticide solution (62 mL) was applied to the soil within the plastic buckets to approximate the 4 liter/1 square meter (1 gal/10 ft^2). The solutions were applied by using a compressed air paint sprayer. Lids were placed on each bucket to prevent evaporation. After 24 hours, a 7.6 ID × 15-cm plastic pipe was pushed into the center of the treated soil, which minimized edge effects caused by the plastic buckets. The pipe was capped, then the bucket was upturned and the soil was allowed to fall out of the bucket but remain in the pipe. A 7.6-cm diameter plastic dowel was used to push the soil out of the pipe at 1-cm increments to a depth of 12 cm. Each soil increment was placed in labeled re-sealable plastic bags. The active ingredients were extracted from the soil and analyzed by procedures described below.

Imidacloprid was extracted and analyzed by a method modified from Peterson (*20*). Recovered soil (15 ± 1 g) was placed in a foil weigh boat and air dried at room temperature overnight. Dried soil (10 ± 0.5 g) was placed in a glass jar and 20 mL of 80: 20 acetonitrile: water solution was added and then the soil was then shaken for 4 hours at 200 rpm. The jars settled for > 48 hours, the liquid was decanted and vacuum filtered through glass fiber filters. The collected filtrate was analyzed for imidacloprid content on a Waters Alliance 2695 liquid chromatograph, consisting of 20 μL injection, water + acetonitrile

(60 + 40 by volume) mobile phase at 1 mL min^{-1} through a Whatman Partisphere RTF C-18 column (4.6 × 250 mm) fitted with an Agilent XDB C-18 (4.6 × 12.5 mm) guard column and UV detection (270 nm) on a Waters 996 photodiode array detector. Percentage recoveries for this method were 89, 95, 99 and 115% at 100 μg/g soil and 89, 91, 102 and 112% at 10 μg/g soil for H, U, P and D soils, respectively.

Fipronil was extracted by placing 35 ± 1 g recovered soil into a foil weigh boat and oven drying at 90 °C overnight. After cooling, 25 ± 0.01 g dried soil was extracted by using a Dionex ASE 200 accelerated solvent extractor. In this method, 60 mL of 70: 30 acetonitrile: acetone mixture is passed through the 25-g sample at 100 °C and 10342 kPa (1500 psi). The sample was concentrated to 10 mL under a nitrogen stream, and the resulting extract was analyzed by an Agilent 6890 gas chromatograph. Each injection was 1 μL. The injector temperature was 250 °C with an Agilent 1909 1A-112 ultra 1 methyl siloxane 25 m × 320 μm inside diameter × 0.52 μm film thickness column, with helium carrier gas at 20 mL/min. The oven temperature program was 50 °C for 1 min, ramped at 30 °C per minute to 200 °C and held for 10 minutes, ramped again by 30 °C per minute to 230 °C and held for 8 minutes, for a total run time of 25 min. An electron capture detector was used at 250 °C. There was a three-minute equilibration time between runs with two needles washes of hexane followed by two needle washes of acetone. Percentage recovery of fipronil by using this method was 113% for H soil, 98.3% for U soil, and 98.2% for D soil at 20 μg/g soil.

A split-plot arrangement was used in a randomized complete block design (blocked by soil type, with all treatments for a particular soil conducted on the same day), with each container (combination of soil moisture and compound) as the whole plot factor and soil depth as the subplot factor. The study had three replications. Mixed analysis of variance on SAS (*21*) was used to determine significance due to soil type, soil moisture and depth.

Termite bioassays

Stock solutions of Premise and Termidor were prepared by serial dilution. For the range finding assay, solutions were prepared so that the compounds were tested at 100, 50, 10, 1 and 0.1 μg/g soil. Each soil was separately treated by adding 10 mL of the appropriate dilution to 100 ± 0.1 g oven-dried soil in plastic bags. The soil was mixed thoroughly and allowed to sit overnight. Three 15-g portions were removed and placed in separate 15 × 60-mm ID Petri dishes. A square of cardboard, 1 × 1 cm, was placed in the dish and ten *Reticulitermes flavipes* workers were added. Survival of termites was counted in each dish at 7 days. Following the range finding assays, fipronil solutions were made to constitute 1, 0.8, 0.6, 0.4, 0.2 and 0.1 μg/g soil and imidacloprid solutions were made to constitute 10, 8, 6, 4, 2 and 1 μg/g soil. The solutions were applied as described above and termites from the same colony used for the range finding test were used in the manner described above. The LC_{50} values and 95% fiducial limits were calculated by using Probit analysis on SAS (*21*).

Results and Discussion

Depth of penetration

Data were not collected for either compound at 15% soil moisture in U soil. This soil saturates at about 16% moisture (Table 1) and standing fluid was observed on the soil surface 24 hours after application.

The effects of soil moisture on concentration were examined for each combination of soil type and compound (Figures 1 and 2). For all soil types, the effect of soil moisture on fipronil concentration depended upon depth; i.e. there was a statistically significant interaction between soil moisture and depth ($P < 0.0001$ for each soil at 14, 42 degrees of freedom for D, H and P and 7, 28 degrees of freedom for U). Fipronil concentration in the top 1 cm declined with increasing soil moisture, while fipronil concentrations at 2 to 5 cm were higher in soil of 10% moisture (Figure 1). Except for D soil at 15% soil moisture, there were no differences in fipronil concentration below 5 cm for any soil type or soil moisture level.

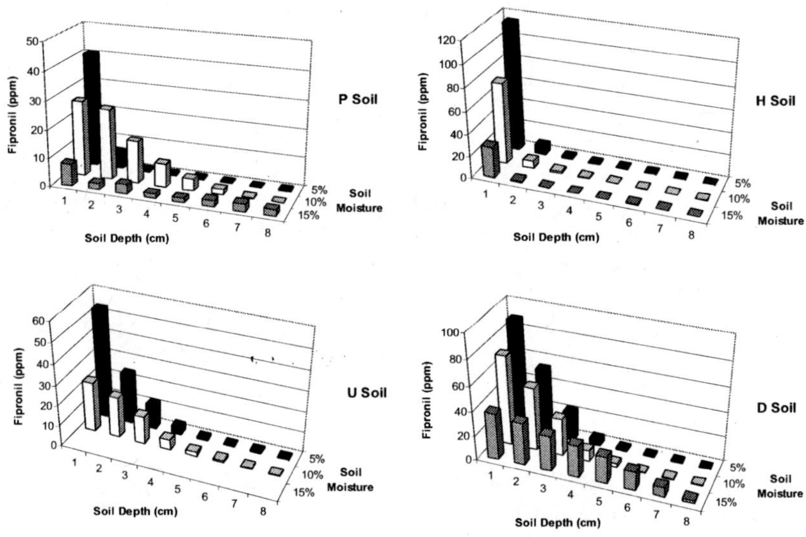

Figure 1. Fipronil recovered from each soil at each depth and moisture combination.

A pattern similar to that observed for fipronil was observed for imidacloprid (Figure 2). There was a significant interaction between soil moisture and depth for U, H and P soils ($P < 0.0001$ at 14, 42 degrees of freedom for H and P, 7 and 28 degrees of freedom for U), but depth was the only significant factor for D soil ($P < 0.0001$ at 7, 42 degrees of freedom). Similar to fipronil, the concentration of imidacloprid in the top 1 cm declined with an increase in soil moisture, although an increase was observed in P soil at 10% soil moisture. Imidacloprid concentration at 2 to 5 cm was higher for 10% soil moisture in P soil, and it was roughly equivalent among the three soil moistures in the other three soils.

Soil types were compared within compounds. For both imidacloprid and fipronil, there was a significant three-way interaction between soil type, soil moisture and depth (imidacloprid: df = 35, 147; $F = 5.56$, $P < 0.0001$; fipronil: df = 35, 154; $F = 7.22$, $P < 0.0001$). Fipronil concentrations were much lower in P and U soils than in H and D soils and the effects due to soil moisture and depth are discussed above. Imidacloprid concentrations were roughly equivalent

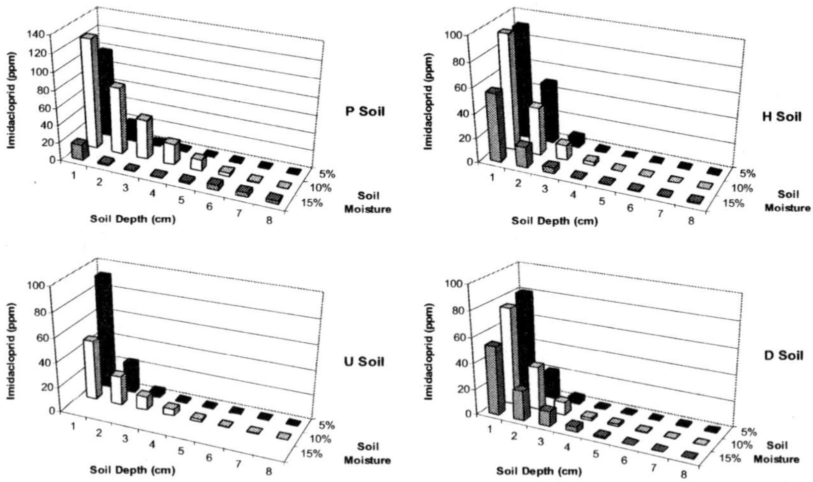

Figure 2. Imidacloprid recovered from each soil at each depth and moisture combination.

between the four soils, but with differences between P and U soils at 10% soil moisture.

Statistical interaction aside, the difference in the penetration of either compound in the soil types is not great. Neither compound penetrated much beyond 5 cm regardless of soil type or moisture, with the exception of fipronil in D soil at 15% soil moisture. Fipronil penetrated the least well into H soil, which is surprising because H soil is relatively sandy. Imidacloprid penetrated similarly into all four soils. Soil type, then, should not be a major factor affecting the initial penetration of a termiticide. Realistically, unless the local building codes require that fill dirt be brought in, soil type is not a choice and even then the fill will more likely be chosen due to expansion and settling potential than for properties conducive to termite control.

Termite bioassays

The LC_{50} values for imidacloprid and fipronil in each of the four soils are shown in Table 2. Both compounds were the most toxic in P soil, which is used as a construction fill and is therefore the most relevant for termite control beneath structures. Fipronil was of equivalent toxicity in U, H and P soils, but less toxic in D soil. D soil had the highest organic matter content of the four soils used, as well as the highest silt, clay and cation exchange capacity. Imidacloprid was equally toxic in all soil types except P soil, where it was more toxic.

Table 2. Seven-day LC_{50} values (95% FL) of fipronil and imidacloprid in µg/g soil applied to the soils used in this study to *R. flavipes*

Soil Type	Fipronil LC_{50} (95% FL)	Imidacloprid LC_{50} (95% FL)
P	0.14 (0.11, 0.17)	3.25 (1.85, 5.18)
D	0.62 (0.48, 0.75)	9.29 (6.39, 18.34)
H	0.18 (0.14, 0.23)	15.11 (12.82, 17.44)
U	0.18 (0.15, 0.21)	12.39 (10.68, 14.14)

Both compounds were the most toxic in P soil, which had the lowest organic matter. A recent study by Mulrooney and Gerard (*22*) found that among four soil types, fipronil was most toxic in a sandy loam soil, followed by sand, a loamy sand and a silt loam, and this trend generally followed a pattern of increasing organic matter. The same general pattern was observed here. Therefore, termiticide-treated soils lower in organic matter should provide the most toxic barrier to termites.

The depth of penetration determines the thickness of the chemical barrier. From these results, it seems that 10% soil moisture for P soil and 15% soil moisture for D soil would provide the thickest barrier. It is noteworthy, however, that there is a reduction in concentration in the top 1 cm with an increase in soil moisture.

This begs the question of what type of barrier is desirable? A thick barrier will withstand minor disturbances more than a thin barrier, but a thick barrier, with lower initial concentration, may degrade to below effective levels more quickly than a thin barrier with higher initial concentration. Figure 3 illustrates this with a hypothetical compound with a half-life of 6 years and that is not effective below 20 µg/g soil. If the barrier is thin, say 1 cm, and the initial concentration of this compound were 100 µg/g soil in the soil, it would take about 14 years to degrade to below 20 µg/g soil. If the initial barrier is thicker, say 3 cm instead of 1 cm, the initial concentration would be lower, here starting at about 60 µg/g soil. In this situation, the barrier would degrade to below 20 µg/g soil in 10 years instead of the 14 years required for the thinner barrier.

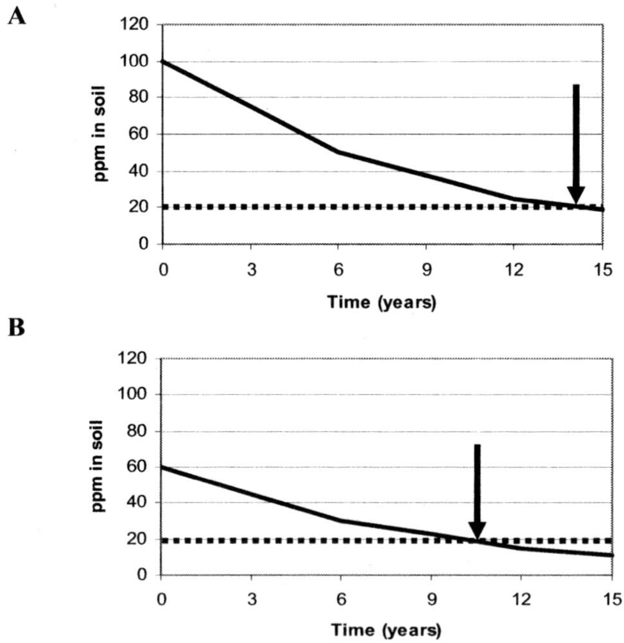

Figure 3. Degradation of a hypothetical compound with a half-life of 6 years and minimum effective concentration of 20 μg/g soil (dashed line). Arrow indicates time when treatment is no longer effective for A) a 1-cm barrier of high initial concentration and B) a 3-cm barrier of lower initial concentration.

Further studies are currently underway to determine how application volume affects the initial thickness of the barrier. Longer-term field studies are necessary to determine how chemical barriers of different thickness affect structural protection.

References

1. Kowal, R. J. and St. George, R. A. Preliminary results of termite soil-poisoning tests. *Journal of Economic Entomology* **1948,** *41,* 112-113.
2. Snyder, T. E. *"White ants" as pests in the United States and methods of preventing their damage*. Farmer's Bulletin #759, United States Department of Agriculture, Washington, DC. 1916, 20 pp.
3. St. George, R. A. Testing soil poisons for termite control. *Pest Control* **1952,** *20(4),* 36.
4. Snyder, T. E. *Injury to buildings by termites.* Leaflet # 101, United States Department of Agriculture, Washington, DC. 1933.

5. Peterson, C. J., Wagner, T. L., Mulroney, J. E., and Shelton, T. G. *Subterranean termites – their prevention and control in buildings.* Home and Garden Bulletin #64, United States Department of Agriculture, Forest Service, Washington, DC. 2006. 32 pp.
6. Smith, M. W. New approaches to "sub" treatment of slab houses. *Pest Control* **1957,** *25(7),* 36, 38, 40.
7. Kowal, R. J. What the USDA is doing about research on termite control in slab construction. Pest Control **1954,** *22(2),* 12, 14, 16, 18.
8. Smith, M. W. Where are we going in our control methods for subterranean termites? *Pest Control* **1956,** *24(11),* 36, 38, 40.
9. Federal Housing Administration. *Minimum property standards for one and two living units*. FHA #300, November 1, 1958.
10. Beal, R. H. and Carter, F-L. Initial soil penetration by insecticide emulsions used for subterranean termite control. *Journal of Economic Entomology* **1968,** *61,* 380-383.
11. Carter, F-L. and Stringer, C. A. Soil moisture and soil type influence initial penetration by organochlorine insecticides. *Bulletin of Environmental Contamination and Toxicology* **1970,** *5,* 422-428.
12. Carter, F-L., Stringer, C. A. and Beal, R. H. Penetration and persistence of soil insecticides used for termite control. *Pest Control* **1970,** *38(10)*, 18-24, 62.
13. Carter, F-L. and Stringer, C. A. Soil persistence of termite insecticides. *Pest Control* **1971,** *39(2)*, 13-14, 16, 18, 29.
14. Beal, R. H. and Howard, R. W. Subterranean termite control: Results of long term tests. *International Biodeterioration Bulletin* **1982,** *18,* 13-18.
15. St. George, R. A. Tests with new insecticides for termite control. *Pest Control* **1952,** *20(2),* 20.
16. Chapman Chemical Company. *Aldrec Emulsifiable Concentrate* [product specimen label]. Chapman Chemical Company, Memphis, TN. 1968 1 pp.
17. Velsicol Chemical Corporation. *Gold Crest Termide Emulsifiable Concentrate* [product specimen label]. Velsicol Chemical Corporation, Chicago, IL. 1983, 1 pp.
18. Budavari, S., O'Neil, M. J., Smith, A., Heckelman, P. E., and Kinneary, J. F. The Merck Index, 12th ed. Merck and Co., Whitehouse Station, NJ. 1996. 1741 pp.
19. Cassel, D. K., and Nielsen, D. R. Field capacity and available water capacity. In A. Klute (Ed.) *Methods of Soil Analysis, Part 1: Physical and Mineralogical Methods* (2nd Edition). Madison, WI: Soil Science Society of America. 1986, pp. 901-926.
20. Peterson, C. J. Mobility and longevity of imidacloprid in soil columns at a termiticidal application rate. *Pest Management Science* **2007,** *63,* 1124-1132.
21. SAS Institute, 2001. *SAS System for Windows*, version 8.02. SAS Institute, Cary, NC.
22. Mulrooney, J. E. and Gerard, P. D. Toxicity of fipronil in Mississippi soil types against Reticulitermes flavipes (Isoptera: Rhinotermitidae). *Sociobiology* **2007,** *50*, 63-70.

Chapter 9

Fipronil: Toxicity to Subterranean Termites and Dissipation in Soils

J.E. Mulrooney[1], T. L. Wagner[1], and P. D. Gerard[2]

[1]USDA Forest Service, Starkville, MS 39759
[2]Clemson University, Clemson, SC

Fipronil (Termidor 80 WG) was applied to covered and exposed plots at one secondary and four primary USDA Forest Service termiticide test sites in the U.S. Residue analyses and bioassays of soil samples were conducted over 5 y. Fipronil had an exponential decay at all sites. The DT_{50} of fipronil in a silt loam soil in Oktibbeha Co., MS in covered and exposed plots was 202 and 177 d, respectively. Fipronil dissipation appeared to be faster at the secondary site (Oktibbeha Co, MS) compared to the primary tests sites. Dissipation was faster in covered plots in AZ and MS compared to FL and SC; while that in exposed plots was similar among sites. In 7 d bioassays, termite penetration of soil cores from primary test sites was significantly greater in exposed plots (33.9 ± 1.2 mm) than that from covered plots (25.7 ± 1.2 mm). Differences in termite penetration and termite mortality between covered and exposed plots at the secondary test site were not significant. Average distance penetrated by termites, averaged over treatments and primary sites, significantly increased during the last three years (37.2 mm) of sampling compared to the first three sampling times (20.7 mm). Termite mortality averaged over sites and years for covered and exposed plots was not significantly different. Freundlich adsorption coefficients (K_f), determined from adsorption isotherms, ranged from 0.14 on a gravelly sand (Pima Co., AZ) to 5.47 on a silt loam (Oktibbeha Co. MS).

Fipronil, a halogen-substituted thioether containing phenylpyrazole insecticide, was developed by Rhone Poulenc in 1987. It acts as an agonist at the gamma-aminobutyric acid (GABA)-gated chloride channel/ionopore complex, and possesses a high level of toxicity to insects because of its specificity to this target site (*1*). Laboratory and field studies conducted by Rhone Poulenc on different soil types under different temperature conditions identified five principal metabolites (desufinyl, sulfone, sulfide, amide, and a photodegradation product), which occur through degradation pathways of hydrolysis, photolysis, oxidation, and reduction (*2*). A comprehensive review of the environmental fate and toxicology of fipronil can be found in Gunasekara et al. (*3*).

The behavior of a pesticide in the soil and its dissipation in the environment are dependent on its adsorption, which in turn depends on the physical-chemical properties of the pesticide, the climate, and the nature of the soil. Adsorption processes control the availability of pesticides for adsorption by plant roots or soil organisms, and their leaching through soil (*4*). Therefore, adsorption is a major influence on the balance between pesticide efficacy and leaching to groundwater. It has also been shown that adsorption limits the degradation of pesticides by reducing their partitioning into the soil liquid phase (*5*). Bobe et al. (*6*) studied the adsorption of fipronil to soils varying in organic matter. Their results showed that adsorption to soil decreased with decreases in organic matter content of the soil. The effect of organic matter content of soil on adsorption of fipronil was also demonstrated by Mulrooney and Gerard (*7*). In contact bioassays of fipronil treated soils, termite mortality (LC_{50}) decreased as organic matter in the soil increased. For example, LC_{50}'s ranged from 0.49 ppm on sandy loam soil with low organic matter (1.8 %) to 6.99 ppm on a silt loam with higher organic matter (2.6 %). Increased adsorption of fipronil to soil with higher organic matter content decreased the amount of fipronil available for transfer to termites.

The movement and degradation of fipronil were investigated in Australian soils following standard termiticide treatment methods (surface application under slab and trenching treatments along walls). Surface application studies in three field sites showed slow dissipation and little movement for fipronil in all three soils under the simulated slab during a three-year period. The greatest mass of the chemical residues remained in the quartzite sand layer (thickness, 5 cm), and only small amounts of these were found to have migrated into the soil layers (depth, 0 -15 cm) underneath the quartzite sand layer. Of the three metabolites (desulfinyl, sulfide, and sulfone) found in soils, the sulfone derivative had the highest concentration. One year trenching studies at two sites in Adelaide, Australia, showed that vertical movement and dissipation of fipronil occurred in the soils. The average concentration of fipronil in the trenches (depth, 0-30 cm) decreased from 33.7 to 14.9 mg/kg in the loam soil at one site and from 39.4 to 14.6 mg/kg in the clay soil from the other site over the year (*8*). Ying and Kookana (*9*) determined the sorption of fipronil and its two main metabolites, desulfynil and sulfide, on a range of soils from South Australia. The Freundlich sorption coefficient (K_f) values, a measure of the relative adsorption capacity of soil, for fiponil on these soils ranged from 1.94 to 4.84. The metabolites had a higher sorption to soils, with K_f values ranging from 11.09 to 23.49 for the sulfide derivative and from 4.70 to 11.77 for the desulfynil derivative. The sorption coefficients of fipronil and its metabolites were found to be better related to soil organic carbon than clay content.

In a study on open field behavior of fipronil under Sahelian conditions (Niger, Africa), the amide and trifluoromethlypyrazole derivative were the principal degradation products recovered from the soil (*10*). This study also included a measure of the mobility of fipronil in soils and showed that fipronil did not migrate below the first 10 cm.

The toxicity of fipronil and its metabolites to several insect species has been investigated. Fipronil sulfone is the major metabolite of fipronil in Southern armyworm larvae and presumably in other insects (*11*). Desulfinyl fipronil, a significant contributor to the effectiveness of fipronil, is the principal photoproduct on plants and soils and is as potent as or more potent than fipronil in toxicity to houseflies (*12*). Mulrooney and Goli (*13*) in topical applications of fipronil and its metabolites determined the order of toxicity (LD_{50}) to boll weevils (*Anthonomus grandis*) to be: sulfide > fipronil > sulfone > desulfinyl.

Fipronil (Temidor 80 WG) was registered for use as a termiticide in September 1999 and became available for use in pre- and post-construction applications in 2000. Unlike the pyrethroid termiticides which are repellent to termites, termites can not detect fipronil and thus enter treated areas where they are poisoned. The toxicity of fipronil to subterranean termites has been documented by several researchers. Ibrahim et al. (*14*) determined the 72 h LD_{50} of fipronil to be 1.36 ng/insect in topical bioassays using *Coptotermes formosanus* Shiraki. Osbrink et al. (*15*) determined the LT_{50} of fipronil to *Reticulitermes virginicus* to be an average of 271 min when termite workers were placed on filter paper treated with 630.65 $\mu g/cm^2$ of fipronil. Remmen and Su (*16*) obtained an LC_{50} of 0.04 ppm after *R. flavipes* workers were exposed to fipronil treated sand for 1 wk. Shelton and Grace (*17*) in a simple donor-recipient test exposed *C. formosanus* workers to sand treated with fipronil at 1, 10, and 100 ppm for 1 h. Mean mortalities of termite donors after 14 d were 36, 36, and 98%, respectively.

Ibrahim et al. (*14*) found fipronil to be repellent to *C. formosanus* at 0.125%. However, Remmen and Su (*16*) observed that fipronil concentrations as high as 64 ppm did not repel *R. flavipes* (Kollar) and *C. formosanus* termites. They observed 89% mortality at 1 ppm and failure of termites to completely penetrate treated sand. They concluded that 1 ppm fipronil may provide an adequate barrier for both *R. flavipes* and *C. formosanus.*

Hu (*18*) found 100% mortality of eastern subterranean and Formosan termites within 3 d at treatment concentrations of 50 and 100 ppm and after 28 d at 1 ppm. Her results also showed that penetration into 50 mm thicknesses of treated sand decreased with increasing concentration. Penetration failure was due to rapid mortality, rather than repellency of fipronil.

Although the toxicity of fipronil to termites is well documented, long term studies of the degradation and toxicity of fipronil applied at termiticidal rates have not been reported. The purpose of this study was to determine the dissipation of fipronil in different soils found at U.S. Forest Service test sites and its efficacy against subterranean termites in laboratory bioassays.

Materials and Methods

Test Sites

The four primary Forest Service termiticide test sites are located in Pima Co., Arizona; Calhoun Co., Florida; Harrison Co., Mississippi; and Union Co., South Carolina. The test site in Arizona is on the Santa Rita Experimental Range managed by the University of Arizona near Greenvalley, AZ. The other sites are in the Chipola (near Panama City, FL), the Harrison (near Saucier, MS), and the Calhoun (near Union, SC) Experimental Forests. These sites represent semiarid (Arizona), temperate (South Carolina), and subtropical climates (Florida and Mississippi). Soil pH ranges from approximately neutral (6.9, Arizona) to moderately acidic (4.8, Florida) (Table 1). In 2001, a secondary test site was established in Mississippi State University's John W. Starr Memorial Forest in Oktibbeha Co. just outside Starkville, MS.

Experiment 1

An approximate area of 10 by 15 m was cleared of small trees and shrubs in the Forest Service's secondary test site in Oktibbeha Co., Mississippi. A 6 by 8 grid consisting of 1.5 M square plots was then measured and marked. Treatments consisting of fipronil (Termidor 80 WG) at 0.06% A.I and water controls in covered and exposed plots were randomly assigned to plots. There were five replicates of each treatment. An approximate 60 by 60-cm area in each plot was cleared of vegetation and duff to expose the mineral soil. A 43 by 43-cm metal treating frame was placed on the soil and rocks and roots in the upper 7 cm of soil were removed. The 0.06% fipronil solution (764 ml volume) was applied within the treating frame using a watering can. This volume corresponds to a standard 3.785 L per 9.29 m^2 (1 gal. per 10 sq. ft.) pretreatment application of termiticide. After treatment, black plastic sheets and 40.6-cm square concrete stepping stones were placed over fipronil and water-only control plots, while five plots of each treatment were left exposed. The design was completely randomized with five replicates of fipronil and water controls in both covered and exposed plots, for a total of 20 plots. Because of the close proximity of this site to the laboratory, soil samples were collected for residue analysis immediately after application and at monthly intervals for the first year, thereafter at yearly intervals for 5 y. Samples were collected at 0 and 6 mo, and at yearly intervals for bioassay. Samples were collected using a sampling probe in which the soil core was collected in 2.54-o.d. by 10.16-cm butyrate (Tenite®, U.S. Plastics, Lima, OH) tubes.

Table 1. Test site characteristics

Site	*Location*	*Soil Type*	*Bulk Density* (g/cm^3)	*pH*	*% OM*	*% Clay*	*% Silt*	*% Sand*	*Ave. Yearly Rainfall (cm)*
Santa Rita Experimental Range	Pima Co., AZ	Gravelly sand	1.47	6.9	0.6	7.5	15.1	77.4	35.5
Chipola Experimental Forest	Calhoun Co., FL	Sand	1.16	4.8	0.8	2.8	2.7	94.5	163
Harrison Experimental Forest	Harrison Co. MS	Sandy loam	1.11	5.1	1.8	4.9	25.2	69.9	170
John Starr Memorial Forest	Oktibbeha Co., MS	Silt loam	0.97	4.8	2.6	15.0	72.2	12.8	130
Calhoun Experimental Forest	Union Co., SC	Sandy loam	1.09	4.7	1.3	3.8	29.0	67.2	127

Experiment 2

Application of a 0.06% A.I. rate of fipronil (Termador 80 WG) in a volume of 3.785 L per 9.29 m^2 was made to soil at the four primary termiticide test sites maintained by the Forest Service.

At each site, plots were laid out in completely randomized design on 1.5 M centers in a 2 by 10 arrangement. In this test, the soil was prepared and fipronil (0.06%) was applied to covered and exposed plots in the same manner as described above.

Installation of each test and site visits each year were made in February, April, June, and September to Florida, Arizona, Mississippi, and South Carolina, respectively. Immediately after application and each following year for 5 y, three soil samples were collected from each plot. Two of the samples were used for bioassays, the third for residue analysis. Soil samples collected from untreated soil were used as controls. Soil samples were held at -20°C until bioassays and residue analyses were conducted.

Residue Analysis

Analysis of fipronil residue was done using an Agilent® (Santa Clara, CA) 5990 gas chromatograph equipped with electron capture detector. The parameters of the analysis method were as follows: injection volume, 1µl; carrier gas, helium; make-up gas, argon/methane; injector temperature, 250°C; detector temperature, 250°C; oven program, 50°C initial temperature with a 30°C/min ramp to 230°C for 8 min. An Agilent® 25-m Ultra-1 methyl siloxane phase column (I.D. 0.32 mm) with 0.52-µm film thickness was used. Retention time was 17.924 min.

Sampling tubes were emptied of soil (~45.9 cm^3); the soil core was mixed and oven dried at 100°C. Then 25 g samples were randomly collected for extraction. All solvents used in extractions were HPLC grade. Extraction of fipronil from soil was made with an Accelerated Solvent Extractor, ASE-200 (Dionex®, Sunnyvale, CA) using a 70:30 mixture of acetone:acetonitrile at a total volume of 50 ml. Oven temperature and pressure were 100°C and 105.4 kg/cm^2, respectively, with a 5 min static time. Extraction volume was reduced to 10 ml under nitrogen using a Rapid Vac (Labconco®, Kansas City, MO). Percent recoveries of fipronil (Termidor 80 WG) spiked in soils from the different sites were: Pima Co. AZ, 108.2 ± 2.4; Calhoun Co., FL 94.5 ± 11.2; Harrison Co., MS; 113.3± 1.9; Oktibbeha Co., MS, 98.2 ± 8.7; and Union, Co., SC, 98.3 ± 7.5.

Adsorption

Adsorption isotherms were obtained using a batch equilibrium method (*6, 9*). Two grams of soil from each test site was treated with 5 ml of 5% acetonitrile/water solutions (0 – 10 ppm) of technical fipronil (98%) (Chem Service, Inc, West Chester, PA) in 20-ml scintillation vials. Vials were shaken in a shaker/water bath for 4 h at 200 rpm and 22°C. After centrifugation at 2800 rpm for 30 min, the supernatant (3 ml) was separated and passed through solid phase extraction (C_{18}) cartridges (AccuBond^{II}, Agilent Technologies Inc., United Kingdom). The cartridge was first conditioned with 5 ml of acetonitrile followed by 5 ml of distilled water before 3 ml of the supernatant was loaded. Elution of fipronil was obtained with 5 ml of acetonitrile. The eluate was brought to dryness under a constant stream of nitrogen and then re-dissolved in 1 ml toluene. Fipronil concentration was then determined by GC-ECD as described above. Percent recovery of fipronil from eluate was 89.4 ± 2.8%.

The amount of fipronil adsorbed was evaluated as the difference between that initially present in the solution and that remaining after equilibration with soil. The adsorption isotherms were obtained by plotting the equilibrium content of fipronil adsorbed to soil against the equilibrium concentration of fipronil in the liquid phase. These isotherm data were described by the Freundlich equation:

$$S = K_f C^n \qquad \text{(equation 1)}$$

Where S is the concentration of fipronil adsorbed by the soil (µg/g), C is the equilibrium concentration ((µg/ml). Values of the parameters of sorption, K_f (Freundlich coefficient) and n (Freundlich exponent), were estimated by linear regression after log-log transformation.

Bioassays

Two 10.0-cm deep soil samples from each plot were bioassayed with termites from two *Reticulitermes* spp colonies. Termites were collected from fallen pine logs separated from each other by at least 1000 m on the Noxubee National Wildlife Refuge near Starkville, MS and held at ambient temperature in galvanized trashcans in the laboratory. Different colonies were used each year.

The bioassay method used was similar to that described by Su et al. (*19*). In our bioassay, the 10.0 cm of soil in the sample tube was reduced to 5.0 cm by pushing out the bottom 5.0 cm of soil. Two 3.0-cm agar segments were placed on either side of the soil core to provide moisture during the bioassay. Then the tube containing the 5.0 cm of soil was connected by a Tygon® tubing collar to another tube containing 80 workers and one soldier. Wooden sticks of southern yellow pine and paper strips provided food and harborage for termites in both the tube containing termites and the tube with soil, so that termites had a source of food both above and below the treated soil.

The bioassay was terminated after 7 d when mortality as well as distance tunneled through treated soil (penetration) was determined.

Data Analysis

The experimental design of both experiments was completely randomized. Distance penetrated by termites into soil cores and termite mortality, adjusted for control mortality (*20*), were analyzed using the mixed procedure (PROC MIXED) of SAS (*21*). Mean separation was made using the PDIFF option.

Results and Discussion

Residue Analysis

Experiment 1

Fipronil applied to sandy loam soil in the John W. Starr Memorial Forest in Oktibbeha Co., MS showed an exponential decay over the five years of the study (Figure 1). Time to 50% dissipation (DT_{50}) in covered and exposed plots was 202 and 177 d, respectively. These values are within the half-life range, 91 – 222 d, in soil determined by Rhone Poulenc (*2*). Fipronil residues in exposed plots leveled off at 2.11 ppm after 12 mo, while those in covered plots did not level off until 24 mo when levels in the soil were 0.79 ppm.

A study conducted between 1990 and 2002 at the Harrison Co. test site determined the half-lives of termiticides applied at label rates to soil in trenches around miniature foundations. These data are presented to give some perspective to the dissipation of fipronil: chlorpyrifos 1.0%, Dursban® TC (1,254 d); fenvalerate 0.5%, Tribute® (831 d); permethrin 0.5%, Dragnet® FT (768 d); cypermethrin 0.3%, Prevail® FT (488 d); cypermethrin 0.25%, Demon® TC (399 d); isofenfos 0.75%, Pryfon 6 (301 d); and permethrin 0.5%, Torpedo® (138 d) (*22*). Fipronil at about one tenth the application rate (0.06%) dissipated slightly slower than the Torpedo formulation of permethrin (0.5%).

Experiment 2

It was not possible to determine DT_{50}'s in this experiment because samples were collected at yearly intervals and 50% of the residue had dissipated by the time the 1 y samples were collected. As in Experiment 1, the dissipation of fipronil at primary test sites in Pima Co., Arizona; Calhoun Co., Florida; Harrison Co., Mississippi; and Union Co., South Carolina appears exponential (Figure 2). Parameters of the regressions are given in Table 2. All regressions were significant ($P<0.0001$). The initial residues collected from all primary test sites were higher than that at the secondary test site in Oktibbeha Co., MS (Experiment 1). The dissipation of fipronil in soils at the primary test sites in Arizona, Florida, Mississippi, and South Carolina does not appear to be as rapid as that in the soil in Experiment 1 from Oktibbeha Co., MS (Figures 1 and 2). The silt loam soil from Oktibbeha Co., MS had higher organic matter (OM) than the other soils in this study. Soils with higher OM could be expected to have higher populations of microbes to hasten the degradation of fipronil.

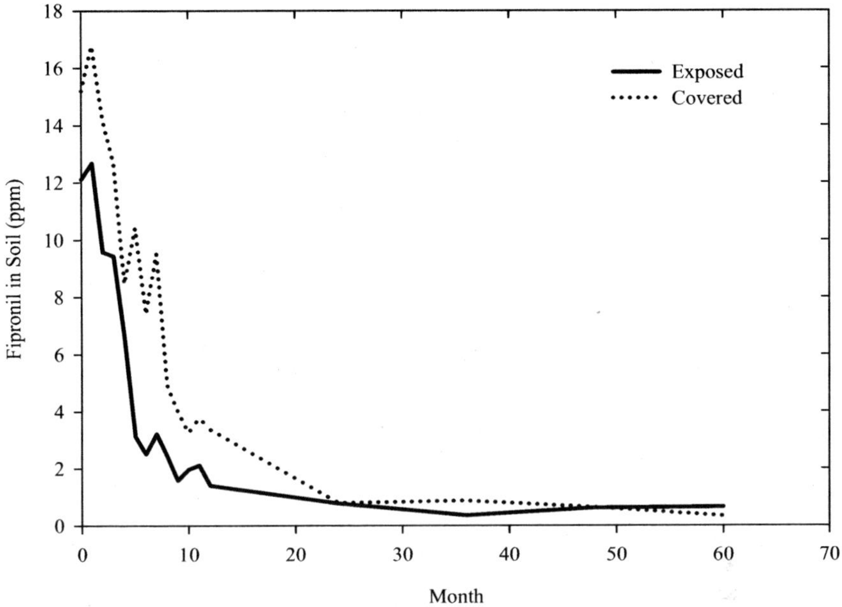

Figure 1. Fipronil residues extracted from covered and exposed plots at U.S. Forest Service secondary test site (Oktibbeha Co., MS).

Degradation of fipronil by microbes in a clay loam soil was demonstrated by Zhu et al. (*23*). They showed that the half-lives of fipronil in a non-sterile clay loam soil were 9.72 and 8.78 d at 25 and 35°C, respectively compared to 33.51 and 32.07 at 25 and 35°C, respectively in the sterile soil. This study demonstrated that microbial degradation was an important factor for the metabolism of fipronil in the non-sterile clay loam soil.

Residues in covered plots did not level off until around 4 y after application, when residues in Florida and South Carolina were about twice those in Arizona and Mississippi (Figure 2). Increased dissipation in Arizona could be the result of high temperatures; while Mississippi's high rainfall and warm temperatures produce a favorable environment for microbes. Residues in exposed plots in Arizona, Florida, and Mississippi leveled off 1 y after application. Those in exposed plots in South Carolina were about ten times those from the other sites after 1 y and did not level off until 3 y after application.

Initial (year 0) amounts of fipronil recovered from soil samples varied among sites. In Experiment 1, 12 to 16 ppm of fipronil were found in the silt loam cores from Oktibbeha Co. collected at time 0 (Figure 1). The amount of fipronil recovered could be a result of the penetration of the fipronil solutions into the soil at application. As will be discussed below, fipronil readily adsorbed to the silt loam in Oktibehha Co. in Experiment 1; therefore, the penetration of fipronil into the soil would likely be very shallow. Fipronil residue would be limited to the upper portion of soil samples (10.0 cm) collected after application.

Because fipronil was not uniformly distributed within the soil sample, a dilution effect occurred as only a portion of the residue was collected when the 25 g of soil was randomly taken from the sample for residue analysis. The amount of fipronil relative to the amount of soil collected is small; therefore low concentrations in the soil were observed. A similar situation existed at the Pima Co. site in Arizona, in that penetration was limited by application to a dry soil in April. Higher initial concentrations of fipronil in soil were found in sand in Calhoun Co., FL and sandy loam in Union Co., SC (Figure 2). The application of fipronil at the Florida site was made in February, a time of year when the soil would be expected to have a high moisture content which would aid in the penetration of fipronil though the soil (*24*). When samples were collected after application at this site, the amount of fipronil relative to the amount of soil collected was high, therefore higher concentrations of fipronil were observed. Also, Carter and Stringer (*25*), in laboratory studies of penetration of chlorinated hydrocarbon termiticides through soils from seven states, observed greater penetration of sand and sandy loam soils from the Florida and South Carolina test sites.

Table 2. Parameters of linear regressions of year (log) on fipronil residue (log) in soil in covered and exposed plots for each Forest Service test site

Site	*Intercept*	*Slope*	r^2
	Covered		
Pima Co., AZ	2.86	-2.26	0.85
Calhoun Co., FL	3.76	-1.21	0.51
Harrison Co., MS	3.64	-2.41	0.73
Union Co., SC	3.66	-1.22	0.39
	Exposed		
Pima Co., AZ	2.78	-2.90	0.83
Calhoun Co., FL	2.91	-2.54	0.84
Harrison Co., MS	2.53	-2.12	0.52
Union Co., SC	3.52	-2.19	0.67

Adsorption

An adsorption isotherm, which describes the relation between the activity or equilibrium concentration of the adsorptive (fipronil) and the quantity of adsorbate (fipronil solution) on the soil surface at constant temperature, is generally used to describe adsorption (*26*).

The parameters of the Freundlich equation (equation 1) are given in Table 3. S is the amount of adsorbed fipronil, C is the equilibrium concentration of dissolved fipronil and K_f and n are two constants characteristic of the fipronil adsorption capacity (*26*). K_f is the amount of fipronil adsorbed at an equilibrium concentration, which is a measure of the relative adsorption capacity of soil and n is the intensity factor of the adsorption (*6*).

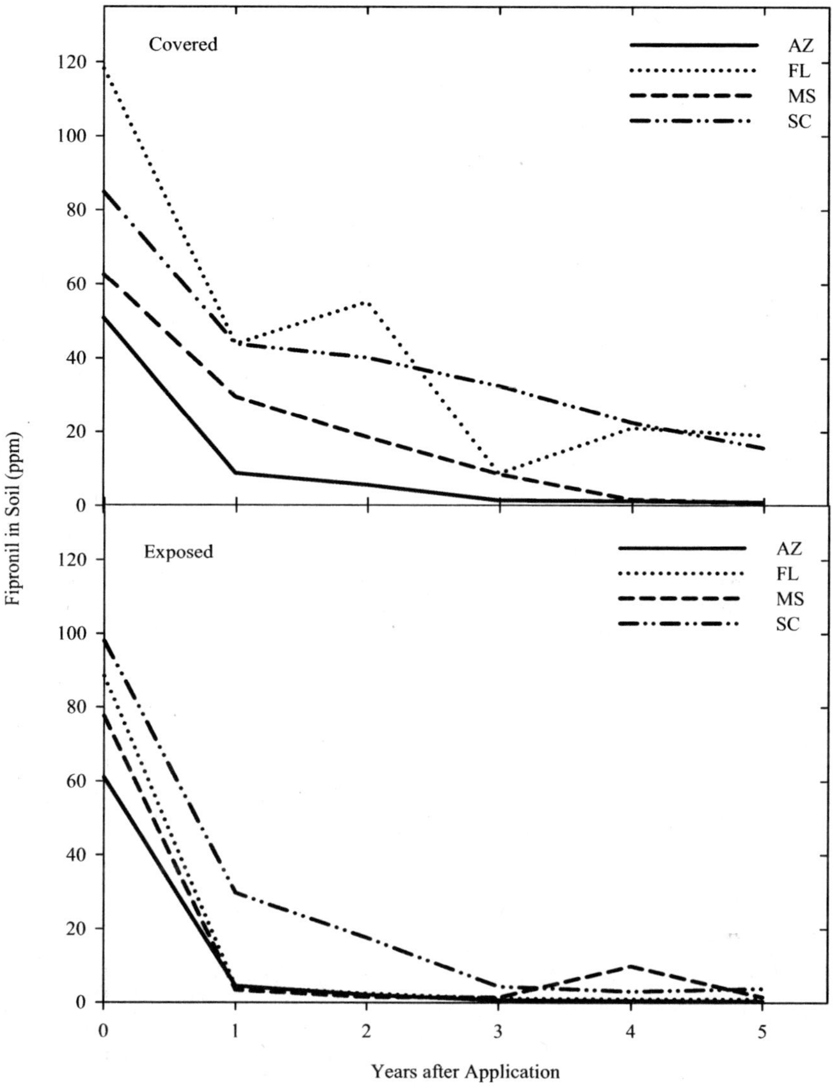

Figure 2. Fipronil residues extracted from covered and exposed plots at U. S. Forest Service primary test sites

Table 3. Parameters of the Freundlich equation ($S = K_f C^n$) describing the adsorption isotherms of fipronil in soils at Forest Service termiticide test sites

Site	K_f	*n*	r^2
Pima Co., AZ	0.14 ± 0.02	1.73 ± 0.21	0.87
Oktibbeha Co., MS	4.82 ± 0.34	1.08 ± 0.19	0.76
Calhoun Co., FL	0.15 ± 0.02	1.47 ± 0.25	0.78
Harrison Co., MS	1.01 ± 0.04	0.93 ± 0.13	0.83
Union Co., SC	0.88 ± 0.08	1.02 ± 0.13	0.85

The silt loam soil from Oktibbeha Co., MS had the highest adsorption coefficient (5.47). Sandy loam soils from Harrison Co., MS and Union Co., SC had coefficients that were similar; 1.21 and 1.20 respectively, but were five times lower than that of the silt loam from Oktibbeha Co. The lighter soils from Pima Co., AZ (gravely sand) and Calhoun Co., FL, (sand) had coefficients much lower than the other soils, 0.14 and 0.15 respectively. As can be seen from Table 2, the soil with the greatest K_f, the silt loam from Oktibbeha Co., had the highest percent OM. This soil also has the highest clay content of the soils included in the test. Soil OM has been shown to be highly correlated with pesticide adsorption. For example, Ying and Kookana (*9*) reported that Freundlich sorption coefficients (K_f) of fipronil on a range of soils from South Australia ranged from 1.94 to 4.84 and were better related to soil OM than clay content. Bobe et al. (*6*) in another study of fipronil adsorption on two Sahelian soils (Saguia and Banizoumbou) from Niger, Africa and a Mediterranean soil (Montpellier) also determined that adsorption was dependent on OM: the adsorption coefficients were 4.3 (Saguia 0.1% OM), 7.3 (Banizoumbou 0.3% OM) and 45.5 (Montpellier 6.5% OM). For unknown reasons, sorption coefficients observed for soils from Pima Co. and Calhoun Co. (0.14 ± 0.02 and 0.15 ± 0.02, respectively) in our study, which had lower percent OM than that of the Saguia soil, were much lower than that of the Saguia soil in Bobe et al. (*6*). Ahmad et al (*27*) reported that the nature of OM is a determining factor of the adsorption capacity of a soil. They found that variation in adsorption of pesticides could be explained only when variations in the aromatic components of OM were taken into consideration. More than likely, the components of OM in soils from Arizona and South Carolina are different from those found in Niger, Africa; to what extent is beyond the scope of the present study.

Bioassays

Experiment 1

Average distance penetrated by termites through 50-mm untreated soil cores was 45.1 ± 2.2 and 48.9 ± 1.5 mm for covered and exposed plots, respectively. Differences in distance penetrated by termites and termite mortality between covered and exposed plots treated with fipronil were not significant; therefore, penetration and mortality were averaged over test type (Table 4). Distance penetrated through soil samples and termite mortality had similar trends in that both increased over time since application. Increases in termite penetration through termiticide treated soil over time should be expected as residues dissipate; however, mortality generally decreases over time due to dissipation. One possible explanation for these results is that for the first 2 y fipronil remained in the upper portion of the soil core and termites only penetrated the core to the edge of the fipronil residue (ca. 24 mm). Over time, fipronil migrated down into the soil, became less concentrated, and thus termites were able to penetrate greater distances into the core. Also, it is possible that roots growing through treated plots and/or excavations by other soil invertebrates create guides and/or passage ways through the soil over time that prompt termites to penetrate through treated soil.

Table 4. Average mortality and distance penetrated through silt loam soil cores (covered and exposed plots combined) treated with fipronil (Termidor 80 WG) in Oktibbeha Co. MS

Month	*Mortality (%)*	*Distance (mm)*
0	68.1 ± 7.8 b	22.2 ± 5.7 b
6	63.6 ± 8.9 b	25.8 ± 5.5 b
12	72.4 ± 6.9 b	46.6 ± 3.6 a
24	89.3 ± 4.7 a	26.8 ± 6.0 b
36	95.1 ± 2.9 a	52.0 ± 0.0 a
48	96.6 ± 2.8 a	42.1 ± 4.6 a
60	86.9 ± 5.9 a	51.8 ± 0.2 a

Means within a column followed by the same letter are not significantly ($P < 0.05$) . different as determined by PDIFF (*21*)

As fipronil became more dispersed in the soil and penetration increased, termite mortality significantly ($F = 5.19$; df = 6, 133; $P < 0.0001$) increased (Table 2). A slight, but non-significant, decrease in mortality occurred at 60 mo after application. Mortality at 24 mo (89.3 ± 4.7%) and thereafter was significantly greater than that at 0, 6, and 12 mo; 68.1 ± 7.8, 63.6 ± 8.9, and 72.4 ± 6.9%, respectively. Control mortality in covered (30.1 ± 3.6 %) and exposed (37.6 ± 3.8%) plots was unexpectedly high in this soil. The abrasiveness of the clay in this soil could have contributed to the mortality of controls. Smith and Rust (*28*) observed that a 2 h exposure of termites dry sand containing 10% kaolin clay resulted in 34% mortality.

Experiment 2

Averaged over years and sites, termites penetrated an average of 49.6 ± 0.6 mm out of a possible 50.0 mm through control soil samples with an overall average mortality of 17.2 ± 1.6%. Distance penetrated through fipronil treated soil samples from covered plots (25.7 ± 1.2 mm) was significantly less ($F = 46.06$; df = 1, 752; $P < 0.0001$) than those from exposed plots (33.9 ± 1.2 mm).

Average distance penetrated by termites through fipronil treated soil samples each year, averaged over treatments and sites, significantly ($F = 54.86$; df = 5, 904; $P < 0.0001$) increased during the last three years (37.2 mm) of sampling compared to the first three sampling times (20.7 mm). This is similar to the results from Experiment 1 and provides additional evidence that downward movement of fipronil may have occurred over time.

Termite penetration through soil samples from the four sites, averaged over treatments and years, was significantly different ($F = 37.76$; df = 3, 4; $P = 0.0032$). Average termite penetration of fipronil treated soil followed the order: Mississippi (36.6 ± 1.4 mm) > South Carolina (35.0 ± 1.3 mm) > Florida (28.3 ± 1.6 mm) > Arizona (15.8 ± 1.4 mm).

The interaction of years*sites*treatment was significant ($F = 3.74$; df = 12, 752; $P < 0.0001$) for distance penetrated by termites through soil samples. Unlike the results of Experiment 1, there were no consistent trends of penetration distances with time since application (Table 5). Average soil penetrations for each year*site combination included the entire gamut of possibilities, 0 to 50 mm, during the 5 y of the study.

Termite mortality in samples from covered and exposed plots in penetration bioassays averaged over years and sites was not significantly different. Mortality for years and sites was significant, $F = 19.03$; df = 5, 896; $P < 0.0001$ and $F = 3.71$; df = 3, 298; $P = 0.0121$, respectively. Termite mortality on soil samples collected during the second year (73.8 ± 3.0%) was significantly lower than that seen during the other years of the study which ranged from 87.6 ± 1.8 to 93.3 ± 1.3% and mortality on samples collected at year 0 (87.6 ± 1.8%) was significantly different from year one (93.3 ± 1.3%). Termite mortality, averaged over years and treatments, on samples from Arizona (76.8 ± 2.1%) and Mississippi (89.1 ± 1.4%) was significantly lower than that from Florida (90.8 ± 1.6%) and South Carolina (94.7 ± 0.9%). These mortalities are somewhat reflective of the lower fipronil residues recovered from soil samples from Arizona and Mississippi compared to those from Florida and South Carolina (Figure 2).

The three-way interaction of years*sites*treatments was significant ($F = 2.28$; df = 15, 896; $P = 0.0035$) for mortality (Table 6). As was the case with penetration, there were no consistent trends of termite mortality over time since application. Except for an uncharacteristic low mortality on soil collected from exposed plots 2 y after application in Arizona (18.2 ± 7.1%), mortalities were between 76 and 100%.

Fipronil is a slow-acting nonrepellent termiticide. The lethal time to kill 50% of the population (LT_{50}) of *Reticulitermes virginicus* was determined for fipronil by Osbrink et al. (*15*). Termite workers placed on filter paper treated with 630.65 µg/cm^2 of fipronil had an average LT_{50} of 271 min compared to an average LT_{50} of 13 min for workers exposed to 526.13 µg/cm^2 of chlorpyrifos, a fast-acting organophosphate termiticide. Fipronil's lack of repellence was demonstrated by Remmen and Su (*16*). They observed that fipronil concentrations as high as 64 ppm did not repel *R. flavipes* or *C. formosanus*

Table 5. Average distance (mm) penetrated by termites through soil treated with fipronil (Termidor 80 WG) at U.S. Forest Service termiticide test sites

Year	*Arizona*	*Florida*	*Mississippi*	*South Carolina*
		Covered		
0	10.1 ± 2.9 Bc	13.0 ± 3.5 Cc	33.2 ± 4.0 Bb	45.0 ± 3.3 Aa
1	1.9 ± 1.9 Bb	7.2 ± 3.9 Cb	5.2 ± 2.8 Cb	30.8 ± 3.9 Ba
2	0.8 ± 0.8 Bc	7.8 ± 4.3 Cc	46.8 ± 3.6 Aa	18.8 ± 5.0 Cb
3	10.4 ± 4.2 Bc	30.2 ± 5.0 Bb	33.4 ± 4.4 Bb	44.9 ± 3.6 Aa
4	9.0 ± 2.9 Bc	20.6 ± 5.0 Bb	52.0 ± 0.0 Aa	20.6 ± 3.1 Cb
5	34.9 ± 5.0 Ab	47.9 ± 2.0 Aa	50.5 ± 1.5 Aa	41.0 ± 4.3 Aab
		Exposed		
0	0.0 ± 0.0 Bb	9.8 ± 3.0 Db	39.2 ± 3.4 BCa	42.9 ± 3.6 Aa
1	4.0 ± 1.5 Bb	33.0 ± 5.6 BCa	6.2 ± 3.0 Cb	44.8 ± 1.9 Aa
2	2.9 ± 1.5 Bb	27.5 ± 5.8 Ca	37.2 ± 5.1 Ba	28.4 ± 5.7 Ba
3	41.6 ± 4.2 Aa	39.0 ± 5.2 Ba	48.2 ± 2.7 Aa	50.2 ± 0.9 Aa
4	38.1 ± 4.8 Ab	52.0 ± 0.0 Aa	51.6 ± 0.4 Aa	24.3 ± 4.4 Bb
5	35.6 ± 5.4 Ab	52.0 ± 0.0 Aa	36.2 ± 4.9 Bb	28.2 ± 5.4 Bb

Means within a treatment (covered or exposed) and within a column (sites) followed by the same upper case letter are not significantly ($P < 0.05$) different as determined by PDIFF (*21*). Means within a treatment (covered or exposed) and within a row (years) followed by the same lower case letter are not significantly ($P < 0.05$) different as determined by PDIFF (*21*) termites and that mortality at 1 ppm was high (89%) indicating that lack of penetration into the treated sand was due to mortality.

Because fipronil is a relatively slow acting termiticide, termites entering treated areas would not die immediately but have time to leave treated areas before death and therefore may transfer fipronil to nestmates through grooming and trophalaxis. These characteristics of fipronil, slow-activity and ability to be transferred, could possibly explain some of the variability between the degree of termite penetration and the resultant mortality seen in experiment 2. Mortality would not only be due to the amount of toxicant taken up by termites as they penetrate treated soil, but could also be due to the amount of fipronil transferred during social interaction between nestmates exposed to fipronil residues and those that never entered treated soil. In addition, the year-to-year variability in mortality and soil penetration could also be due to differences in colonies used in bioassays.

Protection of structures is dependent upon the presence of an effective termiticide barrier to termite attack. Fipronil has been the most effective new termiticide in Forest Service field tests in recent years. It has been 100% effective for 13 y against termite attack in small plot studies at the lowest label rate of 0.06% at all primary Forest Service test sites (*29*). Even though fipronil, applied at the lowest label rate (0.06%), had a faster rate of dissipation than that reported for older termiticide chemistries applied at rates ranging from 0.25 to 1.0%, fipronil's toxicity showed little decrease over the 5 y of this study. This is most likely due to fipronil's toxicity to termites at very low dosages.

Acknowledgements

We thank Craig Bell and Don Fye for assisting in the installation of this study and for collecting soil samples. We also thank Blossie Boyd for conducting residue analyses and laboratory bioassays.

References

1. Gant, D. B., A. E. Chalmers, M. A. Wolf, H. B. Hoffman, and D. F. Bushey. *Rev. Toxicol.* **1998,** *2*, 147-156.
2. Anonymous. 1996. *Fipronil: Worldwide Technical Bulletin.* Rhone Poulenc. Research Triangle Park, NC.
3. Gunasekara, A. S., T. Trung, K. S. Goh, F. Spurlock, R. S. Tjeerdema. *J. Pest. Sci.* **2007,** *32*, 189-199.
4. Coquet, Y. *Pest. Manag. Sci.* **2002,** *58*, 69-87.
5. Guo, L., W. A. Jury, R. J. Wagnet, and M. Flury. *J. Contam. Hydrol.* **2000,***43*, 45-62.
6. Bobe, A., C. M. Coste, and J. F. Cooper. *J. Agric. Food Chem.* **1997,** *45*, 4861-4865.
7. Mulrooney, J. E., and P. D. Gerard. *Sociobiol.* **2007,** *50*, 63-70.
8. Ying, G. and R. S. Kookana. *Environ. Toxicol. Chem.* **2006,** *25*, 2045-2050.

9. Ying, G. and R. S. Kookana. *J. Envion. Sci. Health.* **2001,** *B36,* 545-558.
10. Bobe, A., J. F. Cooper, C. M. Coste, and M. A. Miller. *Pestic. Sci.* **1998,** *52,* 275-281.
11. Brookhart, G. and D. F. Bushey. 1994. *Eight International Union of Pure and Applied Chemistry International Congress of Pesticide Chemistry*, July 4-9 1994, Washington, DC, abstract 189.
12. Hainzl, D., L. M. Cole, and J. E. Casida. *Chem. Res. Toxicol.* **1998,** *11,* 1529-1535.
13. Mulrooney, J. E. and D. Goli. *J. Econ. Entomol.* **1999,** *92,* 1364-1368.
14. Ibrahim, S. A., G. Henderson, and H. Fei. *J. Econ. Entomol.* **2003,** *92*, 461-467.
15. Osbrink, W. L. A., A. R. Lax, and R. J. Brenner. *J. Econ. Entomol.* **2001,** *94,* 1271-1228.
16. Remmen, L. N. and N. Y. Su. *J. Econ. Entomol.* **2005,** *89*, 906-910.
17. Shelton, T. G. and J. K. Grace. *J. Econ. Entomol.* **2003,** *96,* 456-460.
18. Hu, X.P. *J. Econ. Entomol.* **2005,** *98*, 509-517.
19. Su, N. Y., R. H. Scheffrahn, and P. M. Ban. *J. Econ. Entomol.* **1993,** *86*, 772-776.
20. Abbott, W. S. *J. Econ. Entomol.* **1925,** *18*: 265-267.
21. SAS Institute. 2001. *User's Guide*, *ver. 8e*. SAS Institute, Cary, NC.
22. Mulrooney, J. E., T. L. Wagner, B. M. Kard, and P. D. Gerard. *Sociobiol.* **2006,** *48*, 117-133.
23. Zhu, G., H. Wu, J. Guo, and F. M. E. Kimaro. *Water, Air, and Soil Pollution* **2004,** *153*: 35-44.
24. Beal, F. H. and F. L. Carter. *J. Econ. Entomol.* **1968,** *61*, 380-383.
25. Carter, F. L. and C. A. Stringer. *Bull. Environ. Contam. & Tox.* **1970,** *5,* 421-426.
26. Sparks, D. L. *Environmental Soil Chemistry*. Academic Press, Inc. San Diego, CA. 1995.
27. Ahmad, R., R. S. Kookana, A. M. Alston, and J. O. Skjemstad. *Environ. Sci. Tech.* **2001,** *35*, 878-884.
28. Smith, J. L. and M. K. Rust. *J. Econ. Entomol.* **1993,** *86*, 53-60.
29. Wagner, T., T. Shelton, C. Peterson, and J. Mulrooney. *Pest Control.* **2007,** *75*, 58-60, 62, 64, 66-69.

Chapter 10

Least Toxic Strategies for Managing German Cockroaches

Changlu Wang[1] and Gary W. Bennett

Center for Urban and Industrial Pest Management, Department of Entomology, Purdue University, West Lafayette, IN 47907
[1]Current address: Department of Entomology, Rutgers University, 93 Lipman Dr. New Brunswick, NJ 08901

German cockroach resistance development, chronic infestations, and the health impact of insecticide applications have prompted increased interest in least toxic technologies and integrated pest management strategies (IPM) for managing German cockroaches to minimize insecticide use, increase long-term efficacy, and slow down insecticide resistance development. New research data on the relative efficacy of attractants, cockroach pheromones, sticky traps, inorganic insecticides, insect growth regulators, and IPM programs have assisted in the adoption of alternative cockroach management methods. Yet, challenges remain in voluntary adoption of IPM programs and least toxic technologies. Education and coordinated efforts among residents, management staff, pest management professionals, and policy makers will be needed for greater acceptance of least toxic strategies.

Introduction

Of the approximately 4,000 cockroach species in the world, the German cockroach, *Blattella germanica* L., is by far the most successful species. It is found in homes, restaurants, factories, hospitals, ships, and any other indoor environment with food, water, and a warm temperature. Cockroach infestations are often closely related with sanitary conditions in a structure. German cockroaches can reproduce to huge numbers if the room is left untreated. As many as 3,657 cockroaches were trapped in 24 hours on 6 sticky traps placed in one occupied apartment in 2006 (Wang and Bennett, Purdue University, *unpublished data*). Estimated that only 3% of the cockroaches were trapped, the apartment had approximately 122,000 cockroaches!

German cockroaches have significant economic and medical impact. Cockroaches produce allergens and trigger asthma (*1*). In a national survey of 831 U.S. homes, 13% had > 2.0 U/g (units per gram of dust) cockroach allergen (Bla g 1, one of the allergens produced by cockroaches) in kitchen dust samples (*2*). German cockroaches contaminate food and other commodities. The cost for managing German cockroaches can become a significant burden over time, and pesticide use for cockroach management can be a serious health concern for young children.

There are a variety of consumer and professional products for managing German cockroaches, such as sprays, insect foggers, dusts, gel baits, bait stations, and sticky traps. These products vary in cost, formulation, toxicity, and effectiveness. The active ingredients in majority of the insecticide sprays and foggers sold for controlling cockroaches in the U.S. are synthetic pyrethroids. Common active ingredients are allethrin, cyfluthrin, cypermethrin, deltamethrin, lamda-cyhalothrin, permethrin, and resmethrin. Pyrethroids can pose both short and long-term health risks to humans (*3*). Children are at higher risk of pesticide poisoning due to their behavior and developmental stage (*4*). In addition to the pesticide active ingredient, adjuvants such as piperonyl butoxide, which is used to enhance the "knock-down" effect of pyrethroids, and inert ingredients, such as solvents, may cause health problems for sensitive individuals such as children, older adults, and people with chronic illnesses (*5*).

Unaware of the potential health or environmental effects of insecticides, consumers are often guided by price or advertisements in selecting a product. Based on a survey of cockroach control products sold in 106 New York City stores in 2002, insecticide sprays were the most commonly purchased item (*6*). Some residents sprayed daily and used multiple insecticides to kill cockroaches.

Pest management professionals often choose products based on the treatment cost. Periodic applications of gel baits or containerized bait stations containing organic insecticides are the main methods for cockroach management in inner cities by professionals in the U.S. (*7, 8*). Cockroach bait residues are commonly seen in multi-family apartments.

Frequent applications of insecticides not only may pose direct danger to humans, but also induce cockroach resistance and contaminate the environment. High levels of resistance to cockroach baits have been reported (*9-12*). Numerous pesticides have been detected in indoor air and settled dust in homes (*13*). These concerns have led to increased emphasis on alternative strategies and integrated pest management (IPM) for managing cockroach infestations. The objective of this chapter is to review the current status of non-chemical and low-impact chemical methods and IPM for managing German cockroaches.

Cultural Control

Cockroaches need food, water, and harborage to survive. Food residues, unwashed dishes, clutter, and pet food provide favorable conditions for cockroaches to survive and reproduce. Cockroach infestations are closely related to sanitary conditions in the environment and human activities. In an infested apartment, bedrooms usually do not have cockroaches or have the least cockroach numbers simply because there is a lack of food and water in bedrooms.

Cockroaches in a clean, uncluttered environment are much more easily eliminated than those in a dirty environment. Many reports cite the relationship between sanitary conditions and success in cockroach management programs (*14-16*). Reducing or eliminating unsanitary conditions is critical if we are to successfully manage German cockroaches with minimal use of insecticides. Simple practices such as cleaning floors and areas around kitchen appliances, prompt washing of dishes, removing food residues and garbage, and covering open food containers (including pet food) will help reduce cockroach infestations. Reducing clutter is equally important because clutter provides harborage for cockroaches and increases the difficulty in insecticide applications.

Environmental Modification

Cockroaches hide in narrow places. Sealing holes and cracks in the living environment reduces the number of potential hiding places, and improves control efficacy because dust or liquid chemicals are difficult to apply in these places. In multi-family dwellings, cockroaches can migrate between neighboring units through utility penetrations or through doors (*17*). Caulking these areas and adding door swipes will reduce the occurrence of new infestations.

Leaky pipes or faucets provide water sources for cockroaches. In an infested apartment, cockroaches usually concentrate in areas around the kitchen sink, refrigerator, or toilet where water is available (*18*). Prompt repair of malfunctioning faucets, pipes, etc., and cleaning up spills will reduce cockroach survival and reproductive potential.

Physical Control

Trapping

Traps are very useful tools for detecting cockroaches and many other crawling insects in the environment. The following animals were found in the monitoring traps placed in apartments: mice, ants, small flies, spiders, millipedes, and beetles (*18*).

Traps are also helpful for guiding pesticide applications thereby reducing pesticide use. In addition, sticky traps are useful for evaluating insecticide efficacy against cockroaches (*19, 20*). Traps are convenient to use, non-toxic, and inexpensive. Hence, they are frequently used in cockroach management programs.

Trap Types

Two types of traps are used to monitor cockroaches: sticky traps and jar traps. Commercially available sticky traps have many different sizes and shapes. The traps have a thin glue surface to capture cockroaches that wander into the traps. Insect traps are designed to catch crawling insects and other arthropods. Mouse sticky traps can also catch cockroaches. However, mouse sticky trap designs do not have openings on the side or top, which are useful features for catching cockroaches.

Home-made jar traps are made of any type glass jar such as 0.943-liter wide mouth mason jars or 0.124-liter baby food jars (*21, 22*). Food needs to be placed in the jars to attract cockroaches. Bread wetted with beer is most effective in attracting cockroaches into jars. The inner surface of the jar is covered by a thin layer of vaseline and oil to prevent escape. Jar traps are cheaper than sticky traps and are re-usable. However, they are less convenient than sticky traps due to their size and the time needed for preparation.

Use of Traps

Proper placement of traps is very important to maximize trap efficacy. Traps should be placed in locations where cockroaches are likely to hide. In apartments, areas around the refrigerator, stove, kitchen sink, food pantry, and toilet are favored by cockroaches. Traps need to be placed against a wall or a vertical surface (*23*). Depending on the cockroach infestation levels, traps should be checked after one or more nights. In heavily infested areas, a large trap may become full after only one night.

Food and other attractive materials can be placed in the center of most any trap to increase the effectiveness. Wang and Bennett (*24*) studied the effect of five attractants on trap efficiency. All attractants significantly increased the number of cockroaches trapped in sticky traps compared with un-baited traps. Bread with beer was by far the most attractive bait, and increased the trap catches 34-fold over un-baited traps. The bait should be placed on an inverted small bottle cap or any other holding device to avoid wetting the glue surface of the sticky traps. The bait needs to be replaced every 1-2 days to maintain the attractiveness.

Effectiveness of Traps

In general, sticky traps are much more effective than jar traps in catching cockroaches (*24*). The effectiveness of sticky traps varies. Openings on the sides or top of the traps are a helpful feature enhancing the trap efficacy. A smooth surface around the glue area also greatly increases the trap efficiency. Flat glue boards captured significantly more cockroaches than triangular traps in one study (*24*).

Cockroach susceptibility to trapping varies with trap type and cockroach size. Small nymphs are more likely to be trapped by sticky traps than large nymphs. In contrast, jar traps are biased toward large cockroach individuals (adults or large nymphs). Cockroach age structure measured by sticky traps is similar to field population structure, whereas, cockroaches collected from jar traps have a much lower percentage of nymphs than the populations in the natural environment (*24*). Small nymphs might be less likely to climb and/or fall into the jar traps than large nymphs and adults.

Moore and Granovsky (*25*) compared the susceptibility of five cockroach species to trapping. Among them, the Oriental cockroach (*Blatta orientalis* L.) was the easiest to catch and brownbanded cockroach (*Supella longipalpa* (F.)) was least susceptible. A study was conducted in simulated kitchens to compare the efficacy of trapping and baiting against the Oriental cockroach. Sixty Oriental cockroaches were released in each kitchen (5.6 m^2). Ten Trapper insect traps (Bell Laboratories, Inc., Madison, WI, U.S.A.) were placed in one kitchen. The other kitchen was treated with Advion (0.6% indoxacarb) and Maxforce FC Select (0.01% fipronil) cockroach gel baits. After 13 d, the number of Oriental cockroaches decreased by 95% and 100% in the trap and bait-treated kitchens, respectively (Wang and Bennett, Purdue University, *unpublished data*).

Role of Traps in Cockroach Management

German cockroach numbers caught in traps do not change significantly over time at most trap locations (*26*). This feature is useful for estimating population distributions and population changes after pesticide applications. Field studies in apartment buildings showed very consistent distribution patterns. Areas around refrigerators and stoves in the kitchens accounted for 60% of the trap catches (*18*).

Traps can remove a large number of cockroaches when they are placed in multiple locations. In a 29-week study, a median number of 40 traps were placed in 12 cockroach infested apartments; the median number of German cockroaches removed by trapping during the 29-week period was 439 (*27*). Despite the large number of cockroaches that can be removed by traps, traps are not recommended as the sole method for eliminating cockroaches because they are not effective in significantly reducing the cockroach population levels in the living environment (*28, 29*). The most efficient trap (Victor M-330, Woodstream, Lititz, PA, U.S.A.) only trapped an average of 3.7% of the cockroaches per day when $\approx$ 170 cockroaches were present in 1 $\times$ 1 m arenas (*24*).

Vacuuming

Vacuuming immediately removes many cockroaches in heavily infested environments, and removes cast skins, fecal materials, as well as dead cockroaches. These are the allergen sources that can exacerbate asthma. Thus, vacuuming provides additional benefits besides reducing cockroach numbers. Extensive trapping and vacuuming provided a similar level of control of German cockroaches as gel bait in a field trial (*20*). However, a disadvantage of vacuuming is that it takes a much longer time to reduce populations than applying insecticides. In addition, a vacuum with a HEPA filter needs to be used to avoid having allergens blown back into the air.

Electronic Devices

Various electronic devices are advertised for controlling pests. These products are claimed to repel insects and other pests through high frequency sound, electromagnetic waves, or negative ions. Gold (*15*) reviewed scientific studies on some electronic pest control devices and did not find any reports showing the effectiveness of the devices. To date, there are no scientific data proving the effectiveness of electronic devices against cockroaches.

Inorganic Materials

Inorganic materials are some of the oldest insecticides used for controlling cockroaches. They are widely used in managing ants, cockroaches, termites, stored product insects, and other crawling insects. The most common inorganic insecticides are boric acid and other borate materials, formulated as dust, gel bait, or granular bait. The main advantages of inorganic insecticides are long residual activity, low toxicity, no known resistance in cockroaches, and low cost. A comprehensive review of inorganic insecticides used in cockroach management is presented by Ebeling (*30*).

Boric Acid and Other Borate Materials

Boric acid (H_3BO_3), also called boracic acid or orthoboric acid, was first registered as an insecticide in 1948 by the U.S. Environmental Protection Agency for control of cockroaches, termites, fire ants, fleas, silverfish, and many other insects. Among the inorganic insecticides, boric acid is the most commonly used in German cockroach management. A less common borate material is disodium octaborate tetrahydrate.

The mode of action of boric acid against cockroaches is not completely clear. Generally, it acts as a stomach poison affecting the insects' metabolism, and the dry powder is abrasive to the insects' exoskeleton. Ebeling (*30*) suggested that both destruction of the digestive tract and penetration through the exoskeleton contribute to mortality. In a recent study, Habes et al. (*31*) revealed that boric acid infected cockroaches showed destruction of epithelial cells, increased glutathione S-transferases, and lowered acetylcholine esterase activity. This is the first report indicating boric acid dust possesses neurotoxic functions. Symptoms of boric acid poisoning include erratic behavior, tremors and paralysis.

Boric acid is a slow-acting chemical. In choice assays, mortality plateaued after nine days of exposure to boric acid dust on vinyl panels at 1.5 mg/cm^2 (Wang and Bennett, Purdue University, *unpublished data*). Neurotoxin baits (such as fipronil, indoxacarb) can cause 100% mortality within 1-2 days. The slow-acting feature is a main limiting factor to the adoption of boric acid when speed of control is emphasized.

No significant resistance to boric acid in insects has been found. Although field cockroach strains are consistently less susceptible than the laboratory strains based on our experiments, they can be effectively controlled by boric acid dust. Thus, boric acid is still being widely used and effective against many crawling insects. Boric acid dust offers satisfactory control results in residences and commercial facilities at a very low material cost.

Boric acid is generally considered much safer to human and animals than organic insecticides because it has a relatively high LD_{50} (lethal dose to cause 50% mortality of the population) value and it does not volatilize. However, boric acid is used as undiluted or slightly diluted dust. Accidental ingestion, skin contact, and inhalation of boric acid dust can pose health risks to human or animals.

Dust Formulation

Among the various borate materials used for controlling cockroaches, dust is the primary formulation. In choice tests where cockroaches were provided with treated and untreated harborages, German cockroaches did not avoid boric acid treated harborage at its minimum effective rate (0.61 mg/cm^2 or 0.02 oz/ft^2) (Wang and Bennett, Purdue University, *unpublished data*). Significant repellency was found starting from 3.04 mg/cm^2. These results imply that in field applications, a thin layer of dust is more effective and economical than a thick layer of dust, which will repel cockroaches and reduce the efficacy.

In laboratory experiments evaluating the efficacy of boric acid dust applied to harborages, there was a clear concentration-mortality response at the rates below 1.5 mg/cm^2. Beginning from 1.5 mg/cm^2 concentration, boric acid dust caused > 96% control mortality to field strain cockroaches. In a similar experiment, Nibor-D (98% disodium octaborate tetrahydrate) dust (Nisus Corp., Rockford, TN, U.S.A.) applied at 1.5 mg/cm^2 induced 99% mortality at 7 d against the laboratory strain of the German cockroach. Its efficacy against two field strains was much lower, with only 61-70% mortality after 21 d exposure. The LT_{90} (time at which 90% of the population is killed) values against the laboratory strain and two field strains were 6, 27, and 28 days, respectively (*32*).

Field data documenting boric acid efficacy against German cockroaches is scarce. Ebeling et al. (*33*) reported in German cockroach infested apartments, thorough application of boric acid dust at the rate of 454 g per apartment resulted in 99.7% and 100% trap catch reduction after 1 and 3 months, respectively. Moore (*34*) compared boric acid alone and boric acid plus silica dust treatments in apartments. Approximately 227 g dust was applied per apartment. Cockroach counts were reduced to < 4 after 3 months in all treatments. In livestock production systems, the efficacy of boric acid dust was comparable to organic residual insecticide for managing German cockroaches (*35*). High humidity or water in the environment does not affect the toxicity of boric acid dusts and silica gel (*36*). They remain effective after absorbing moisture from the air or wetted by water in the environment.

Boric acid dust involves higher labor cost than baits during application. In addition, boric acid dust cannot be applied into certain places or surfaces (e.g. corners of door frames, door hinges, window shades, metal surface). Thus, in heavily infested areas, additional tools need to be used to achieve satisfactory control.

Solid Bait Formulation

Boric acid baits are generally toxic but have moderate performance in field trials (*37, 38*). Dry or wet bait containing boric acid and disodium octaborate tetrahydrate is repellent to German cockroaches (*39*). This is the main reason that solid baits containing boric acid are not very effective in field studies. Despite the shortcomings, several boric acid gel baits and granular baits are available in the U.S. Two common cockroach baits are InTice roach bait (30% orthoboric acid) and Niban FG granular bait (5.0% orthoboric acid). They have the advantage of easy application, but are much more expensive (3-4 times) than dust formulations. There are no reported data on their effectiveness to field cockroach populations.

In laboratory experiments, LesCo granular bait (5% orthoboric acid; LesCo, Inc. Cleveland, Ohio, U.S.A.) caused 72% mortality to the laboratory German cockroach strain, whereas, only 5-13% mortality occurred among the three field strains (Wang and Bennett, Purdue University, *unpublished data*). In another laboratory test, the Niban-FG granular bait (5% orthoboric acid; Nisus Corp., Rockford, TN, U.S.A.), caused 100% and 65.9% mortality to the laboratory and field strains, respectively (Table I) (*40*).

Table I. Effectiveness of boric acid granular bait (Niban FG) against laboratory and field strains of the German cockroach

Cockroach strain	*LT_{50} (95% FL) (day)*	*LT_{90} (95% FL) (day)*	*Mean control mortality after 28 d*
Jwax (laboratory strain)	3.9 (3.4-4.7)	5.9 (5.1-7.4)	100%
Dorie (Field strain)	24.9 (23.0-27.6)	37.6 (33.7-43.7)	65.9%

In laboratory choice tests using small number of cockroaches (15-20 per box), boric acid gel bait caused near 100% mortality to field strains after 11-21 days of exposure. However, in large mixed-stage population studies (> 200 cockroaches per box), < 50% mortality occurred from boric acid gel bait treatment.

Compared with gel baits containing conventional organic insecticides, boric acid gel bait is much less palatable and effective (*41*). Laboratory assays showed boric acid gel bait was significantly less palatable than cockroach baits containing fipronil (Maxforce FC Select cockroach gel bait), indoxacarb (Advion cockroach gel bait), dinotefuran (Advance cockroach gel bait), and acetamiprid (Transport cockroach gel bait) (Wang and Bennett, Purdue University, *unpublished data*). Field cockroach populations often have multiple food sources and hence, elimination by boric acid gel bait will be very challenging.

Liquid Bait Formulation

Liquid boric acid baits are used for controlling urban ant pests such as Argentine ant, black carpenter ant, Florida carpenter ant, odorous house ant, red imported fire ant, and pharaoh ant (*42-46*). Strong et al. (*39*) showed water based liquid bait containing boric acid or disodium octaborate tetrahydrate was not repellent to German cockroaches. They cause mortality and alter cockroach behavior. However, no liquid boric acid baits are currently available for managing cockroach infestations.

Laboratory assays showed that boric acid was more effective than sodium tetraborate or disodium octaborate tetrahydrate at controlling cockroaches (*47*). Aqueous solutions containing mixtures of 0.5-2% boric acid and any of several inexpensive sugars, including fructose, glucose, maltose, and sucrose as a phagostimulant, at 0.05-1% molar concentrations were effective in controlling field German cockroach populations. A boric acid-based sugar water solution was tested for managing German cockroaches in swine farms (*48*). Bait consisting of 1 or 2% of boric acid and 0.5 M sucrose provided effective population reductions. Cockroach populations were reduced by > 90% within 1-2 months and the populations were maintained below threshold levels. Liquid boric acid bait has not been tested in other commercial facilities or residential structures, Challenges remain in developing a convenient delivery method.

Diatomaceous Earth

Diatomaceous earth (DE) is primarily used to manage stored grain pests (*49*). The mode of action of DE is generally accepted as a desiccation effect on insects. Similar to boric acid materials, DE has very low mammalian toxicity. Nevertheless, health risks of newly developed DE formulations are unclear.

DE deposits are repellent to German cockroaches. There are no reports of population elimination by DE application. This is probably due to the lack of effectiveness in high humidity environment and repellency.

Faulde et al. (*50*) reported the following factors are related to the effectiveness of DE: oil-carrying capacity, humidity, and origin of the DE. The oil-carrying capacity is positively correlated with DE's efficacy against German cockroaches. High humidity decreases the effectiveness of DE. DE formulations based on freshwater diatoms are more effective when compared with those including marine diatoms. Adding hydrophobic silanes can compensate the loss of toxicity at high humidity (> 80% relative humidity) and achieve complete eradication of test populations in simulated field conditions. Further studies on effectiveness and health risks of hydrophobic DE formulations are needed to manage German cockroaches.

Biopesticides

Biopesticides are certain types of pesticides derived from animals, plants, bacteria, and certain minerals. Biopesticides fall into three major classes: microbial pesticides (consist of a microorganism as active ingredient), plant incorporated protectants, and biochemical pesticides. Much of the research related to cockroach management has been on essential oils and fungal pathogens.

Essential Oils

Many essential oils and plant extracts have repellent activity against German cockroaches (*51-54*). They may be useful in protecting sensitive areas and equipment from cockroach infestations.

The volatile components of essential oils can be classified into four main groups: terpenes, benzene derivatives, hydrocarbons, and other miscellaneous compounds (*55*). Essential oils are used in fragrance and flavor industries for producing food flavorings, cosmetics, and detergents. Ngoh et al. (*56*) tested the insecticidal and repellent properties of nine volatile constituents of essential oils against the American cockroach, *Periplaneta americana* (L.). They found the benzene derivatives were more toxic and repellent than terpenes. The LC_{95} (concentration at which 95% of the population is killed) values of eugenol, methyl-eugenol, safrole, and isosafrole were between 0.33-0.46 (mg/cm^2). The KD_{50} (24 h) (concentration at which 50% of the population is knocked down) values were between 0.23-0.49 (mg/cm^2). Safrole was the most effective repellent among the nine tested compounds.

Fungi

Metarhizium anisopliae (Metschnikoff) Sorokin is a well known entomopathogen used for controlling hundreds of insect pests (*57*). Kaakeh et al. (*58*) reported > 90% mortality with *M. anisopliae* strain ESC-1 in controlling German cockroaches by contact method. The same strain induced 50-57% mortality by topical application at a concentration of 4.18×10^8 spores/ml (*59*).

M. anisopliae needs at least three weeks to induce 90% cockroach mortality, which is a serious drawback for its use as a control agent for German cockroaches.

One product, Bio-Path, was registered by the U.S. Environmental Protection Agency for cockroach control in 1993 (EcoScience Corporation, Worcester, MA, U.S.A.). Lack of field efficacy and slow control led to its discontinuation (*60*). Subsequent studies focused on combinations of *M. anisopliae* and chemicals to improve the field efficacy. Kaakeh et al. (*61*) found *M. anisopliae* and an imidacloprid bait had a synergistic effect against the German cockroach. Sublethal doses of chlorpyrifos and cyfluthrin and propetamphos enhanced the effect of *M. anisopliae* in laboratory experiments (*59*). However, in vitro studies indicated adverse effect of chlorpyrifos, propetamphos, and cyfluthrin on the growth and sporulation of *M. anisopliae* (*60*). Zurek et al. (*62*) demonstrated 12.5% boric acid dust or 0.1% (w/v) boric acid in drinking water had synergistic effect to *M. anisopliae*. Boric acid enhances the pathogenic activity of the fungus and not vice versa. The mechanism was suggested as physicochemical or immunologically based.

Ascomycetous fungi in the order Laboulbeniales are known as ectoparasites of millipedes, mites, and insects (*63, 64*). The order contains nearly 2,000 described species worldwide. Among these, about 25 species in the genus *Herpomyces* are parasites of cockroaches (*65, 66*). Infected German cockroaches have shortened and curled antennal flagella, uneven wings, darkened and flaccid cadavers, and putrefied odor. Symptoms develop after 20 days and death occurs within 30 days. There is no field data demonstrating Ascomycetous fungi as an effective biological control agent against German cockroach populations.

Pheromones

Cockroach fecal materials contain aggregation pheromones, which are attractive to German cockroaches (*67, 68*). Crude extract from feces and a mixture of six carboxylic acids are very effective attractants to German cockroaches in laboratory assays (*69*). Cockroach fecal extract can increase food consumption (*70*), efficacy of toxic baits (*71*), and trap catches (*72*). At least one glue trap product contains cockroach pheromones extracted from cockroach feces (Woodstream Corporation, Lititz, PA, U.S.A.). Laboratory studies showed aggregation pheromones can significantly increase trap efficacy (*73*).

Other Organic Non-neurotoxins

Steltenkemp (*74*) described N-monosubstituted neoalkanamides of 11 to 14 carbon atoms for repelling cockroaches, including American, German and Oriental cockroaches. This group of chemicals is also effective against mosquitoes (both *Anopheles* and *Aedes*), black flies and carpenter ants, and to some extent against deer ticks. High levels of repellency to German and American cockroaches by alkyl and aryl neoalkanamides were reported (*75*). These compounds may be useful for repelling cockroaches and preventing cockroach infestations.

Insect Growth Regulators

Insect growth regulators (IGRs) are a group of compounds that disrupt the natural growth and development of insects. IGRs are selective to insect pests and have less effect on other animals than conventional insecticides. IGRs are a desirable alternative for managing German cockroaches because this insect has a relative short life cycle and high reproductive potential. Currently, only juvenile hormone mimics have been registered for control of the German cockroach. They cause sterility of adult cockroaches. Two commonly used IGRs are hydroprene and methoprene. Juvenoid IGRs applied alone provide relatively slow population control against German cockroach and are typically recommended for use with a companion conventional insecticide (*76*).

Another potential group of IGRs for cockroach management is the chitin synthesis inhibitors (CSIs). Flufenoxuron, lufenuron, and noviflumuron are some of the CSIs which have strong population effects against the German cockroach (*77-81*). These chemicals act much faster than juvenoids. CSIs cause mortality of nymphs during molting and adult sterility. Lufenuron sprays caused complete mortality of German cockroach populations after 12 months in simulated domestic environments (*78*). Flufenoxuron wettable powder formulation achieved > 80% population control in multifamily apartments within 8 wk of treatment (*77*). Noviflumuron gel bait caused 95.0% and 97.1% median trap count reduction after 8 and 20 weeks, respectively (Wang and Bennett, Purdue University, *unpublished data*). However, CSIs are not commercially available for cockroach management. The availability of a wide variety of other effective and fast-acting products might have contributed to the lack of interests in developing CSI-based cockroach control products.

Integrated Pest Management

Integrated pest management, or IPM, is a sustainable approach to managing pests by combining biological, cultural, physical, and chemical tools in a way that minimizes economic, health, and environmental risks (*definition from the National IPM Network*). The goal of urban IPM programs is to manage pests by the most economical means, and with the least possible hazard to people, property, and the environment. IPM has gained increased attention in rural and urban settings in recent years. Various residential IPM programs are proposed. In general, they include the following four groups of elements: periodic monitoring, education of property management staff and residents, non-chemical methods (cleaning, reducing clutter, reducing harborages and pest entry points, vacuuming, trapping), and selective use of chemical methods (boric acid, baits, etc.). Interviews with residents and staff, visual inspection, and laying monitoring traps are used during periodic monitoring to understand the pest infestation levels and environmental conditions that contributing to the pest infestations. Staff and resident education includes topics such as cockroach prevention, proper housekeeping, modification of the environment, and record-keeping. Non-chemical methods are used to prevent infestations and eliminate the existing infestations. Heavy cockroach infestations often need chemical tools such as dust or bait to supplement the non-chemical tools. Robinson and Zungoli (*82*) discussed model cockroach IPM programs in various settings.

Kramer et al. (*83*) compared four treatment strategies in apartments. A combination of baiting, cleaning, and education resulted in more rapid reduction in cockroach numbers than baiting alone.

Brenner et al. (*84*) evaluated the effectiveness of an IPM program in East Harlem, New York City, NY. The IPM treatment included education of residents on housekeeping, repairs by a project handyman, fixing plumbing leaks, and providing cockroach bait stations and gel baits to residents. The control group did not receive these treatments. Sticky traps were laid to monitor cockroach infestations. The proportion of intervention households with cockroaches declined from 81 to 39% after six months. By contrast, the control households showed no reduction (from 78 to 81%).

Miller and Meek (*85*) compared the cost and efficacy IPM and insecticide sprays for managing cockroaches in low-income housing. The average cost of IPM was three times of the spray treatment. However, IPM caused 84% cockroach population reduction within four months. The population levels in the spray treatment remained steady for the first five months and increased afterwards during the summer.

Williams et al. (*86*) compared IPM with conventional calendar-based pest control in schools for 12 months. The IPM included initial vacuuming and monthly use of baits and IGR device. The two treatment strategies incurred similar total costs and the efficacy of both treatments was similar. IPM treatment indicated most of the conventional treatments were unnecessary.

Wang and Bennett (*87*) compared IPM and baiting alone for managing German cockroaches in low-income apartments. The IPM treatment included education, vacuuming, trapping, and placing gel baits. The baiting treatment only used gel bait. After 8 months, IPM and baiting resulted in 100 and 95% reduction in trap counts, respectively. The cumulative cost of IPM was $64.5 and $35 per apartment, respectively. IPM treatment resulted in significant improvement in sanitatary conditions of the apartments.

Although the above experimental IPM programs showed various advantages over the chemical-only method, voluntary IPM adoption is very limited. The initial high cost and the need for involvement of multiple parties in education, coordination, and record keeping make it less appealing to property management staff. When selecting a pest control contractor, property managers are compelled to select the lowest bid. Pest management companies often offer low-bids in order to obtain a contract, and the low-bid practice often does not provide effective pest infestation reduction and long-term control.

Conclusion

Sanitation, trapping, vacuuming, dusting, and insect growth regulators are effective tools for reducing/eliminating German cockroach infestations with no or minimum environmental contamination. A combination of several methods should be considered for effective control of cockroach infestations. Using IPM strategies will greatly reduce the insecticide use, improve long-term effectiveness, and reduce cockroach allergens levels.

References

1. Arruda, L. K.; Vailes, L. D.; Ferriani, V. P. L.; Santos, A. B. R.; Pomes, A.; Chapman, M. D. *J. Allergy Clin. Immunol.* **2001,** *107*, 419-428.
2. Cohn R. D.; Arbes S. J. Jr.; Jaramillo, R.; Reid L. H.; Zeldin, D. C. *Environ. Health Perspect.* **2006,** *114*, 522-526.
3. Ray, D. E.; Forshaw, P. J. *Clin. Toxicol.* **2000**, *38*, 95-101
4. Rudel, R. A.; Camann, D. E.; Spengler, J. D.; Korn, L. R.; Brody, J. G. *Environ Sci. Technol.* **2003**, *37*, 4543-4553.
5. Watson, W. A.; Litovitz, T. L.; Rodgers, G. C.; Klein-Schwartz, W.; Youniss, J.; Rose, S. R.; Borys, D.; May, M. E. *American J. Emerg. Med.* **2003**, *21*, 353-421.
6. Carlton, E. J.; Moats, H. L.; Feinberg, M.; Shepard, P.; Garfinkel, R; Whyatt, R.; Evans, D. *J. Comm. Health* **2004**, *29*, 231-244.
7. Greene, A.; Breisch, N. L. *J. Econ. Entomol.* **2002**, *95*, 1-13.
8. Harbison, B.; Kramer, R.; Dorsch, J. *Pest Contr. Tech.* **2003**, *33*(1), 24-29, 83.
9. Morrison, G.; Barile, J.; Macom, T. E. *Pest Contr. Tech.* **2004**, *34*, 62, 64, 66.
10. Wang, C.; Bennett, G. W. *J. Econ. Entomol.* **2004**, *97*, 2067-2072.
11. Liang, D. In *Proceedings of the fifth international conference on urban pests*, Lee, C. Y.; Robinson, W. H. Eds.; P&Y Design Network. Penang, Malaysia, 2005; pp 107-114.
12. Miller, D.; McCoy, T. C. In *Proceedings of the Fifth International Conference on Urban Pests*, Lee, C. Y.; Robinson, W. H. Eds.; P&Y Design Network. Penang, Malaysia, 2005; pp 115-121.
13. Garry, V. F. *Toxicol. Appl. Phamacol.* **2004**, *198*, 152-163.
14. Schal, C. *J. Econ. Entomol.* **1988**, *81*, 536-544.
15. Gold, R. In *Understanding and controlling the German cockroach*; Rust, M. K.; Owens, J. M.; Reierson, D. A. Eds.; Oxford University Press: New York, 1995; pp 325-343.
16. Lee, C. Y.; Lee, L. C. *J. Vector Eco.* **2000**, *25*, 218-221.
17. Owens, J. M.; Bennett, G. W. *J. Econ. Entomol.* **1982**, *75*, 570-573.
18. Wang, C.; Abou El-Nour M.; Bennett, G. W. *J. Comm. Health*, **2008**, *33*, 31-39.
19. Owens, J. M.; Bennett, G. W. *Environ. Entomol.* **1983**, *12*, 1040-1046.
20. Kaakeh, W.; Bennett, G. W. *J. Econ. Entomol.* **1997**, *90*, 976-982.

21. Artyukhina, I. N. *Med. Parazitol. Parazit. Bolezni* **1972**, *41,* 472-477.
22. Reierson, D. A.; Rust, M. K. *Pest Contr.* **1977**, *45 (10)*, 40, 42-44.
23. Owens, J. M. In *Understanding and controlling the German cockroach;* Rust, M. K.; Owens, J. M.; Reierson D. A. Eds.; Oxford University Press: New York, 1995; pp 93-108.
24. Wang, C.; Bennett, G. W. *Environ. Entomol.* **2006**, *35*, 765-770.
25. Moore, W. S.; Granovsky, T. A. *J. Econ. Entomol.* **1983**, *76*, 845-849.
26. Appel, A. *J. Econ. Entomol.* **1998**, *91*, 1136-1141.
27. Wang, C.; Bennett, G. W. *J. Econ. Entomol.* **2006**, *99*, 879-885.
28. Ballard, J. B.; Gold, R. E. *J. Kan. Entomol. Soc.* **1983**, *56*, 506-510.
29. Ballard, J. B.; Gold, R. E. *J. Econ. Entomol.* **1984,** *77*, 661-665.
30. Ebeling, W. In *Understanding and controlling the German cockroach*; Rust, M. K.; Owens, J. M.; Reierson, D. A. Eds.; Oxford University Press: New York, 1995; pp 193-230.
31. Habes, D.; Morakchi, S.; Aribi, N.; Farine, J.-P.; Soltani, N. P*estic. Biochem. Physio*. **2006**, *84*, 17-24.
32. Wang, C.; Bennett, G. W. *Arth. Mgt. Test* **2006,** *30*, L1.
33. Ebeling, W.; Reierson, D. A.; Wagner, R. E. *J. Econ. Entomol.* **1968,** *61*, 751-761.
34. Moore, R. C. *J. Econ. Entomol.* **1972**, *65*, 458-461.
35. Zurek, L.; Gore, J. D.; Stringham, S. M.; Watson, D. W.; Waldvogel, M. G.; Schal, C. *J. Econ. Entomol.* **2003**, *96*, 1362-1366.
36. Appel, A. *J. Econ. Entomol.* **2004**, *97*, 1009-1016.
37. Appel, A. G. *J. Econ. Entomol.* **1990**, *83*, 153-159.
38. Appel, A. G. *J. Econ. Entomol.* **1992**, *85*, 1176-1183.
39. Strong, C. A.; Koehler, K. G.; Patterson, R. S. *J. Econ. Entomol.* **1993,** *86*, 1458-1463.
40. Wang, C.; Bennett, G. W. *Arth. Mgt. Test* **2006,** *30*, L2.
41. Durier, V.; Rivault, C. In *Proceedings of the 3rd International Conference on Urban Pests;* Robinson, W. H.; Rettich, F.; Rambo, G. W. Eds.; Grafické Závody Hronov, Czech Republic, 1999; 113-119.
42. Klotz, J. H.; Moss, J. I. *J. Entomol. Sci.* **1996**, *31*, 9-12.
43. Klotz, J. H.; Oi, D. H.; Vail, K. M.; Williams, D. F. *J. Econ. Entomol.* **1996**, *89*: 673-677.
44. Klotz, J. H.; Vail, K. M.; Williams, D. F. *J. Econ. Entomol.* **1997**, *90*, 488- 491.
45. Klotz, J. H.; Vail, K. M.; Williams, D. F. *J. Econ. Entomol.* **1997**, *90*, 523-526.
46. Klotz, J.; Greenberg, L.; Venn, E. C. *J. Econ. Entomol.* **1998**, *91*, 910-914.
47. Gore, J. C; Schal, C. *J. Econ. Entomol.* **2004**, *97*, 581-587.
48. Gore, J. C.; Zurek, L.; Santangelo, R. G.; Stringham, S. M.; Watson, D. W.; Schal, C. *J. Econ. Entomol.* **2004**, *97*, 715-720.
49. Ebeling, W. *Ann. Rev. Entomol.* **1971**, *16*, 123-158.
50. Faulde, M. K.; Scharninghausen, J. J.; Cavaljuga, S. *J. Stored Prod. Res.* **2006**, *42*, 253-263.
51. Inazuka, S. *J. Pestic. Sci.* **1982**, *7*, 133-143.
52. Inazuka, S. *J. Pestic. Sci.* **1983**, *8*, 293-299.

53. Karr, L. L.; Coats, J. R. *J. Pestic. Sci.*, **1988**, *13*, 287-290.
54. Coats, J. R.; Karr, L. L.; Drewes, C. D. In *Naturally occurring pest bioregulators*. Hedin, P. A. Ed.; ACS Symposium Series #449. American Chemical Society: Washington, DC, 1991; pp 305-316.
55. Haagen-Smit, A. J. In *The Essential Oils*, Guenther, E., Ed.; Van Nostrand Co.: New York, 1948; vol. 1, pp 17-83.
56. Ngoh, S. P.; Choo, L. E. W.; Pang, F. Y.; Huang, Y.; Kini, M. R.; Ho, S. H. *Pestic. Sci.* **1998**, *54*, 261-268.
57. Zimmermann, G. *Pestic. Sci.* **1993**, *37*, 375-379.
58. Kaakeh, W.; Reid, B. L.; Bennett, G. W. *J. Entomol. Sci.* **1996,** *31*, 378-390.
59. Pachamuthu, P.; Kamble, S. T. *J. Econ. Entomol.* **2000**, *93*, 60 – 70.
60. Pachamuthu, P.; Kamble, S. T.; Yuen, G. Y. *J. Econ. Entomol.* **1999**, *92*, 340-346.
61. Kaakeh, W., Reid, B. L.; Bohnert, T. J.; Bennett, G. W. *J. Econ. Entomol.* **1997**, *90*, 473-482.
62. Zurek, L.; Watson, D. W.; Schal, C. *Biol. Contr.* **2002**, *23*, 296-302.
63. Tavares, I. I. In *Insect-Fungus Symbiosis: Nutrition, Mutualism, and Commensalism*; Batra, L. R. Ed.; John Wiley & Sons: New York, 1979; pp 229–258.
64. Tanada, Y.; Kaya, H. K. *Insect Pathology*; Academic Press: San Diego, CA, 1993.
65. Archbold, E. F.; Rust, M. K; Reierson, D. A.; Atkinson, K. D. *Environ. Entomol.* **1986**, *15*, 221-226.
66. Gemeno C.; Zurek, L.; Schal, C. *J. Invertebr. Pathol.* **2004**, *85*, 132-135.
67. Ishii, S.; Kuwahara, Y. *Appl. Entomol. Zool.* **1967**, *2*, 203-217.
68. Rust, M. K.; Appel A. G. *Ann. Entomol. Soc. Am.* **1985**, *78*, 107-110.
69. Scherkenbeck, J. Nentwig, G., Justus, K.; Lenz, J.; Gondol, D. et al. *J. Chem. Ecol.* **1999**, *25*, 1105-1119.
70. Miller, D. M.; Koehler, P. G.; Patterson, R. S. *J. Econ. Entomol.* **1996,** *89, 668-672.*
71. Miller, D. M.; Koehler, P. G.; Patterson, R. S. *J. Econ. Entomol.* **1997**, *90*, 483-487.
72. Miller, D. M.; Koehler, P. G.; Nation, J. L. *J. Econ. Entomol.* **2000**, *93*, 865-870.
73. Kaakeh, W.; Bennett, G. W. *Arth. Mgt. Test.* **1996**, *21*, 392.
74. Steltenkamp, R. J. U.S. Patent 5182305, **1993**.
75. Steltenkamp, R. J.; Hamilton, R. L.; Cooper, R. A.; Schal, C. *J. Med. Entomol.* **1992**, *29*, 141-149.
76. Bennett, G. W.; Reid, B. L. In *Understanding and controlling the German cockroach*; Rust, M. K.; Owens, J. M.; Reierson, D. A. Eds.; Oxford University Press: New York, 1995; pp 267-286.
77. Reid, B. L.; Appel, A.G.; Demark, J. J.; Bennett, G. W. *J. Econ. Entomol.* **1992**, *85*, 1194-1200.
78. Mosson H. J.; Short, J. E.; Schenker, R.; Edwards, J. P. *Pestic. Sci.* **1995**, *45*, 237-246.

79. Ameen, A.; Kaakeh, W.; Wang, C.; Bennett, G. In *Proceedings of the 4th International Conference on Urban Pests*. Jones, S; Zhai, J.; Robinson. W. H., Eds.; Pocahontas Press, Inc.: Blacksburg, VA, 2002; pp. 147-153.
80. Ameen, A.; Wang, C.; Kaakeh, W.; Bennett, G. W.; King, J. E.; Karr, L. L.; Xie, J. *J. Econ. Entomol.* **2006**, *98*, 899-905.
81. Wang, C.; Bennett, G. W. *Pest Manag. Sci.* **2006,** *62*, 434-439.
82. Robinson W. H.; Zungoli, P. A. *In Understanding and controlling the German cockroach*; Rust, M. K.; Owens, J. M.; Reierson, D. A. Eds.; Oxford University Press: New York, 1995; pp 345-359.
83. Kramer, R. D.; Nixon, W. J.; Rosa, R.; Frazier, R. S. *Pest Contr. Tech.* **2000**, *28(5)*, 58, 62, 67-68, 70, 142.
84. Brenner, B. L.; Markowitz, S; Rivera, M.; Romero, H.; Weeks, M.; Sanchez, E. et al. *Environ. Health Persp.* **2003**, *111*, 1649-1653.
85. Miller, D.; F. Meek. *J. Econ. Entomol.* **2004**, *97*, 559-569.
86. Williams, G. M.; Linker H. M.; Waldvogel, M. G.; Leidy, R. B.; Schal, C. *J. Econ. Entomol.* **2005**, *98*, 1275-1283.
87. Wang, C.; Bennett, G. W. *J. Econ. Entomol.* **2006**, *99*, 879-885.

Chapter 11

Movement of Diazinon Residues into Homes Following Applications of a Granular Formulation to Residential Lawns

D. M. Stout II, M. K. Morgan, P. P. Egeghy, J. Xue

United Stated Environmental Protection Agency, National Exposure Research Laboratory, Research Triangle Park, North Carolina C, 27711 USA.

A pilot study was conducted to examine the movement of diazinon following applications of a granular formulation to residential lawns. The objectives included examining the transport and fate of diazinon from an outdoor source to the indoor living areas of six homes, and estimating potential human exposure of six children (age 5-12) living in these homes using site specific data and model default assumptions. Sampling included the collection of the pesticide formulation, soil, particles from doormats, transferable residues from residential turf and indoor flooring, indoor air from living rooms and children's bedrooms, and vacuum dislodgeable dust. Samples were collected from six single family homes located in the Piedmont region of North Carolina between April and August 2001. Environmental samples were collected prior to pesticide application and at days 1, 2, 4, and 8 following the application. Soil concentrations, an indicator of source strength, were highest immediately following the application and declined by an average of 51% by day 8. Transferable residues from turf were determined with the polyurethane foam (PUF) roller and ranged from 0.1 to 970 ng/cm^2 over the study period. Particle-associated residues collected from doormats located at entryways into the home declined from day 2 to day 8 by an average of 75%. Indoor air concentrations in both living rooms and children's bedrooms reached maximal levels from 1 to 2 days following pesticide application and declined over the remainder of the study. Indoor transferable residue levels from carpeted surfaces were typically below the limit of detection and are reflective of a low efficiency collection technique. Concentrations in vacuum dislodgeable dust were variable over time, but consistently exceeded pre-application concentrations. Results demonstrate that the physical translocation of particle-bound residues and the intrusion of volatilized diazinon contribute to indoor levels. Increased airborne concentrations demonstrate the intrusion of diazinon from the outdoor source. Elevated concentrations in dust suggest the movement and deposition of volatilized and/or particle-bound residues. Model estimates suggest that exposure occurred over the duration of the study and that the estimated absorbed mass declined little over 8 days. In summary, the applications to residential lawns resulted in a sustained increase of diazinon levels above background concentrations inside of all homes. Lawn applications were found to be a source of potential occupant exposure both on treated lawns and inside homes.

Introduction

Insecticides are commonly applied to residential lawns, school grounds, parks, golf courses, and athletic fields to control for soil borne insects that can damage turf grass and to control terrestrial insect pests such as ants, ticks, fleas and crickets. In the United States (U.S.) a variety of insecticide formulations, spanning multiple chemical classes, may be conveniently purchased by the general public to control insects, fungi and weeds on their personal lawns. Estimates derived from the home and garden market survey conducted in 2000 to 2001 indicate that about 78 million U.S. households used pesticides, spending nearly 1.3 billion dollars to purchase insecticides and applying 888 million pounds of active ingredient *(1)*. The US EPA *(2)* reported that about 4.5 million kg of diazinon (O, O-diethyl O-2-isopropyl-6-methylpyrimidin-4-yl phosphorothioate) was used in 1981 on residential lawns and turf farms. In 2001, when this study was performed, the same organophosphate insecticide was ranked as the fourth most commonly used pesticide. Due to emerging health and ecological risks, manufacturers of diazinon agreed to a phase out and cancel all residential use products. Effective as of December 31, 2004 no diazinon products with residential uses were to be registered or sold *(3)*. Registrations allowing for use of these compounds in and around homes and buildings have been removed and the products are no longer sold to the general public.

Studies have shown that pesticides applied in residential dwellings move from their point of application and contribute to increased indoor concentrations *(4, 5, 6, 7, 8, 9, 10)*. The movement or translocation of pesticide residues is dependent on the formulation, the physio-chemical properties of the active ingredient(s), the surfaces onto which the applications are made and human activity patterns around the point of application. Pesticide residues are degraded by factors such as heat, hydrolysis, and microbial activity typically found out of doors. Indoors, where residues are protected from these degradation factors, residues may persist, accumulate *(11, 12)* and serve as a continued source of exposure to the occupants. Lewis and MacLeod *(13)* determined that indoor air concentrations of pesticides may be 10 to 100 times higher than those measured out of doors.

Although less studied, applications to residential lawns and foundations of homes have been shown to result in the movement of pesticide residues away from the point of application and into homes. Drift and subsequent volatilization can contaminate outdoor surfaces and contribute to indoor air and surface concentrations. A professional application of microencapsulated cyfluthrin to the exterior perimeter of homes resulted in the deposition of residues up to 9.1 m from the homes foundations and low level contamination on some indoor surfaces *(14)*. Lewis *et al* (7) investigated indoor residues insecticide concentrations over 12 days following the application of an emulsifiable concentrate formulation of chlorpyrifos to the exterior perimeter and foundation of a home. They found that the pesticides applied outdoors were transported indoors and deposited onto surfaces. Research conducted by Nishioka *et al. (15, 16, 17, 18)* substantiated "track-in" as a pathway for the transport of pesticide residues into homes. They showed that foot traffic transported residues from turf treated with the herbicides 2,4-D and dicamba onto carpeting up to 1 week following an application. In a later study, the movement of 2,4-D, chlorpyrifos and dicamba applied as a spray and granular formulation, respectively, to turf was shown to result in increased indoor concentrations. Their findings suggested that rain events and volatilization were important dissipation pathways for these compounds and that "track-in" occurred over the six days following the application. In a third study, 2,4-D was applied to the turf of eleven occupied and two unoccupied homes. Indoor air concentrations and surface loadings were primarily attributed to the re-suspension of residue laden floor dust and subsequent deposition onto surfaces. Results showed that indoor pesticide residues were elevated in the homes with higher occupant activity levels and indoor/outdoor pet dogs, and suggested the residues measured indoors were likely associated with particles tracked-in by the occupants who performed the applications and through the activities of pet dogs. Morgan *et al. (19)* conducted a one home feasibility study that examined the role of a pet dog as a vehicle for transporting diazinon residues following a lawn application. Based on questionnaires and recall diaries it was determined that the dog spent more time on the treated turf relative to the occupants and it was hypothesized that the pet was an important mechanism for the transport of residues into the indoor living area of the home.

In the study presented here, we measured diazinon concentrations from various media following an application of a granular formulation of diazinon to residential turf. It expands on the previously mentioned study *(19)* examining the role of pet dogs as a mechanism for the movement pesticides. The objective of this study was to examine the transport pathway of a semi-volatile insecticide following a granular application to residential turf in six homes, generate input data for human exposure models and to estimate potential human exposure of six children (age 5-12) living in these homes using site-specific data and model default assumptions.

Materials and Methods

From April through August of 2001 six residential homes located in the Piedmont region of North Carolina were monitored following granular applications of diazinon to their residential lawns. Single family households, having one or more adults and children (< 14 years of age) and one indoor-outdoor adult dog and who had previously planned to perform this type application were solicited for participation in the study. Occupants purchased their own commercially-available granular diazinon at local stores and applied the insecticide to their lawns using manually operated rotary spreaders, except for participant 6 who hand broadcast the granular formulation. The applicators were encouraged to read and follow directions as provided on the product labels.

This was a "human observational exposure" study, as defined in 40 CFR Part 26.402. The study protocol and procedures to obtain the assent of the children and informed consent of their parents or guardians were reviewed and approved by an independent institutional review board (IRB) and complied with all applicable requirements of the Common Rule regarding additional protections for children. In addition, the study protocol and procedures were reviewed and approved by an independent Animal Care and Use Committee (ACUC).

Family members recorded in recall diaries their general daily activities. Participant activities were recorded one week prior to the implementation of field sampling and continued on each sampling day throughout the study.

Prior to the application the participant's yards and indoor living areas were measured and graphed on paper. In addition, diagrams of the area treated for each home were prepared prior to the applications to determine the theoretical application rate and actual square meters treated.

Sampling was carried out prior to the application and at 1, 2, 4, and 8 days post-application. Sample types included an aliquot of formulated material collected from the hopper of the spreader or the bag directly, soil cores, turf transferable residues, doormat sweepings at a common occupant/pet entryway, indoor air, vacuum dislodgeable particles (carpet dust) and indoor surface transferable residues from carpeting. Pre- and post-application weights of the formulation were recorded to obtain the total mass applied to the yard.

The aliquots of the granular formulation were placed into 250-mL glass jars using a spatula and nitrile gloved hands. Jars were enclosed in heavy plastic bags to minimize cross contamination with other samples.

Thirty soil samples were collected from thirty different locations within the treated area and composited. The treated area was diagrammed and marked with two bisecting lines that extended to the furthest edges of the treated area. A total of thirty flags designating sample locations were distributed along the two lines. A soil core was collected from within a 30.5 cm radius of each flag at each sampling interval (day). Samples were collected using a T-handle soil probe with a 2.5 cm diameter core. The probe was inserted into the ground to an approximate depth of 1 cm. The soil cores from each home for each sampling interval (day) were aggregated into a 250-mL glass jar capped with a Teflon lined lid.

A single turf transferable residue sample was collected per sampling interval (day) using a polyurethane foam (PUF) roller apparatus and a dry PUF sampling ring *(16, 19, 20)*. The 8-cm wide PUF roller was rolled on the turf at a rate of 10 cm/s over a 2 m distance. Each location sampled was marked on a diagram of the house and lawn and was not re-sampled at any subsequent sampling intervals. Sample locations were within the confines of the area treated by the occupant.

A single doormat sample was collected at days 2, 4, and 8 post-application. Entryway deposits were collected by placing a new, solid black rubber doormat (43 cm X 64 cm) at the door most commonly used to enter and exit the home by the occupants and pets. The entire doormat was vacuumed with a modified hand vacuum fitted with an in-line filter that was hand fabricated from vacuum cleaner bags. The doormat was removed at each sampling interval and replaced by a new unexposed doormat. Vacuuming was accomplished by pulling the intake nozzle over the top doormat surface in a zig-zag pattern across the length of the mat followed by a similar pattern conducted vertically over the mat. Following the systematic vacuuming of the entire top surface of the doormat, the in-line filter was removed and the top folded over to enclose the contents and closed shut with a metal clip. Using gloved hands the filter and its contents were placed into a 250-mL glass jar.

Indoor air was sampled in the room where the participants spent the most time (typically the living room) and in a child's bedroom. Air was sampled from the two locations prior to the application and at 1, 2, 4 and 8 days post application. Samples were collected using low-flow pumps placed in the center of each room *(21)*. Polyurethane plugs (PUF) enclosed in glass housings were connected to the pumps via PTFE tubing. The PUF plugs were open faced with no particle filtration. The sampling heads were suspended 75-cm from the floor with the inlet positioned downwards. Flow rates were set at 3.8 L/min and each sample was collected over 24-h resulting in a sample volume of 5.5 m^3. Following the completion of sampling the glass housings containing the PUF sampling, media were capped on both ends and placed into glass jars.

A high volume small surface sampler (HVS_3) *(22)* was used to collect vacuum-dislodgeable dust from 1 m^2 areas from locations in the carpeted areas of living rooms or dens on days. A single location was selected and sampled prior to the application and on days 2, 4, and 8 post application. To determine transferable residues indoors, PUF roller samples were collected from carpeted surfaces in the living room. Locations sampled were not re-sampled on any later days.

In general all samples were placed in appropriately sized pre-cleaned glass jars with PTFE lined lids. The lids were checked for tightness prior to placing into ice chests for storage at reduced temperatures and under darkened conditions for transport to the laboratory. The chain of custody was established and samples were archived at -20 °C until shipped for chemical analysis. Quality assurance consisted of field duplicates, field spikes, and field blanks, and selected results are shown in Table 1.

Table 1. The limits of detection for spiked matrices, lowest detectable levels for field samples and per cent recoveries for diazinon spiked media.

Matrix	*Matrix Detection Limit*	*Method Detection Limit*	*%Recovery Diazinon*
Air (PUF)	4.7 ng/PUF	0.9 ng/m^{3a}	88
PUF Roller	4.7 ng/PUF	0.003 ng/cm^{2b}	75
Soil	0.2 ng/ dry g	1.7 ng/ dry g	95
Formulation	0.6 ng/g	0.6 ng/g	NA[c]
Doormat	0.2 ng/g	0.2 ng/g	74
Carpet Dust	0.6 ng/g	0.1 ng/g	109

[a] calculated based on average flow rate of 3.8 L/min for 24-h. [b] Calculated based on 20 cm X 200 cm sampling area. [c]NA indicates that values were not determined for the matrix.

The samples collected in this study were generally extracted in an automated accelerated solvent extraction apparatus (Dionex ASE 200™) in 100% hexane (analytical grade) at 100 °C and 2000 psi. The extracts were collected in 60 mL glass tubes and concentrated to a volume of 2 mL in a Turbo Vap concentrator set at 42 °C and 20 psi. Sample cleanup was performed using solid phase extraction tubes (0.5 g Florisil, 3 mL capacity, Restek REPREP™). Tubes were conditioned by flushing with 90:10 hexane:acetone. Samples were eluted with 40 mL of 90:10 hexane:acetone. The eluant was reduced to 1 mL and transferred to a 2 mL vial until chemical analysis. Samples were chemically analyzed using gas chromatography and a mass spectrometriy (GC-MS). The GC oven was programmed to ramp from 100 °C to 150 °C at a rate of 5 °C/min and held for 2 min. The MS was operated at 70 eV using electron impact ionization in selective ion monitoring mode. Calibrations standards ranged from 50 to 200 ng/mL. Laboratory controls included matrix blanks, spikes and spike duplicates and the use of surrogate standards. The quality assurance criterion for spiked matrices and surrogate recoveries was between 40 and 120% and the data were surrogate corrected. All statistical analyses were performed using fixed effects and mixed-effects models (SAS Proc Reg and Proc Mixed procedures) available with SAS version 8.2 (SAS Institute Incorporated, Cary NC). All measurements were log transformed prior to analysis.

The US EPA/ORD/NERL Stochastic Human Exposure and Dose Simulation model for multimedia, multi-pathway pollutants (SHEDS-multimedia version 3.0) *(23, 24)* was used to estimate the exposures through the inhalation, non-dietary ingestion, and dermal routes and subsequent intake for children in this study. Briefly, SHEDS simulates individuals from user specified population cohorts by selecting daily sequential time-location-activity diaries from surveys contained in the US EPA's Consolidated Human Activity Database (CHAD) *(25)* relevant to the specific demographic (age, gender, etc.) characteristics of the participants. Each simulated individual is randomly assigned an appropriate activity diary according to demographic characteristics, and the values for each model input parameter (Table 2) are randomly sampled from distributions and inserted into pathway algorithms. An individual's time series of exposure and absorbed mass by pathway is estimated, and the metric of interest (e.g. time-averaged exposure or absorbed dose) is computed for the individual. The process is repeated thousands of times using Monte Carlo simulation to produce a population distribution of exposure or intake *(26, 27, 28)*. The SHEDS model has the capability to employ pesticide levels measured

Table 2. SHEDS model input parameters

Input Parameter	*Distribution Type*	*Frequency*	*Parameter Estimate*[a]			*Units*	*Reference*
			v_1	v^2	v_3		
Inhalation absorption factor	Triangle	Person	0.5	0.9	1	[-]	*(9, 38)*
Dermal absorption rate for dust or soil	Uniform	Person	0.001	0.3		1/day	*(39)*
Dermal absorption rate for surface residues	Point	Person	0.03			1/day	*(39)*
GI absorption rate for dust or soil	Point	Person	0.8			1/day	*(40, 41)*
GI absorption rate for surface residues	Point	Person	0.8			1/day	*(40, 41)*
Soil-skin adherence factor	Point	Person	0			mg/cm^2	*(24)*
Body-surface fractional contact rate	Beta	Hour	3	6.7		1/hr	*(24)*
Hand-surface fractional contact rate	Beta	Hour	9.4	33		1/hr	*(9)*
Maximum dermal loading for body	Uniform	Person	0.1	0.7		ug/cm^2	*(9)*
Maximum dermal loading for hands	Uniform	Person	0.1	2		ug/cm^2	*(24)*
Fraction of body unclothed	Beta	Day	3	6.7		[-]	*(28)*
Fraction of surface of single hand that enters mouth	Beta	Person	3.7	25		[-]	*(28)*
Dust ingestion rate	Point	Person	1			mg/hr	*(42)*
Soil ingestion rate	Point	Person	1			mg/hr	(24)
Bathing removal efficiency	Beta	Hour	17.1	5.1	0	[-]	(24)
Mouthing removal efficiency	Triangle	Hour	0	0.16	0.32	[-]	(24)
Hand washing removal efficiency	Beta	Hour	32	22		[-]	(9)
Surface-skin transfer coefficient for body (clothed)	Lognormal	Hour	5900	3.8		cm^2/hr	(8, 9)
Surface-skin transfer coefficient for body (unclothed)	Lognormal	Hour	5900	3.8		cm^2/hr	(8, 9)
surface-skin transfer coefficient for hand	Lognormal	Hour	2029	5		cm^2/hr	(9).

[a] distribution (v1, v2, v3); triangle (minimum, mode, maximum,); uniform (minimum, maximum); normal (mean, standard deviation); lognormal (geometric mean, geometric standard deviation).

in exposure media for individuals within a given scenario *(8)*. The model couples activity data with environmental data using physically based equations and calculates resulting exposures and doses.

This modeling application was used to apportion the per cent contribution of each non-dietary pathway to exposure and absorbed mass for each child participant. The daily inhalation time series absorption profile for each individual was estimated as a product of the actual airborne concentrations ($\mu g/m^3$), a basal breathing rate (m^3 air/day), an activity specific ventilation rate ratio, and the duration of the macro-activity event (day/event). The dermal exposure time series dose profile was calculated as a product of the measured surface concentration (soil and turf transferable outdoors and carpet dust indoors; $\mu g/cm^2$), and the dermal transfer coefficients for a given macro-activity (cm^2/h). In addition to mass loading and removal mechanisms such as hand washing, bathing and hand-to-mouth transfers were also considered by the model. The estimated non-dietary ingestion absorption profile included exposure from hand- and object-to-mouth activities. The dermal exposure for the subjects hands (µg) were halved to represent the mass loading to one hand. The mass was adjusted for the fraction of residue on the hand that contacts the mouth per mouthing event and the saliva removal efficiency. The daily absorbed dose (µg/kg/day) was calculated with a simple pharmacokinetic model in SHEDS using appropriate absorption factors for diazinon. Relevant parameters utilized in the model and their estimates are given in Table 2. The participant's physical attributes used in the SHEDS model are provided in Table 3. The contribution from dietary ingestion was disregarded as part of this modeling exercise.

It is important to note that the assumptions shown in Table 2 are based on literature derived inputs and other defined sources. The reader is cautioned that SHEDS model inputs can be changed and those changes might significantly alter the relative importance of the calculated exposure pathways. In addition, the authors assumed that all diazinon residues were equally available for transfer and uptake whether bound to particles or not.

Table 3. Physical attributes of participants that were used as input parameters for the SHEDS model

House #	*Sex*	*Age*	*Weight (kg)*
1	Female	11	26
2	Female	5	21
3	Male	5	17
4	Female	7	28
5	Female	12	34
6	Male	12	36

Results and Discussion

Mass Applied and Area Treated

The homeowners applied the granular diazinon to lawn areas ranging from 139 m^2 (house 3) to 1300 m^2 (house 5) with a mean treated area of 669 m^2 (Table 4). The total amount of granular formulation applied ranged between 2.0

kg (house 1) to 12.7 kg (house 4) with a mean amount of 6.0 kg. The application rates were determined by dividing the mass of formulated material applied by the area treated. The theoretical application rates were based on a labeled rate of 2 lbs/1000 ft^2 or 0.5 g/m^2. Four out of six homes did not exceed the manufacturer's recommended rates and in fact tended to perform applications at rates below those described on the product label. Two homes (3 and 4) did exceed the recommended label rates, by 2 and 6.5 times, respectively.

Table 4. Diazinon concentrations measured from commercially available (5% active ingredient [w/w]) granular formulations.

House	*Total Amount of Formulation Applied (kg)*	*Theoretical Active Ingredient Applied (g)*	*Actual Area Treated (m^2)*	*Calculated Application Rate (g/m^2)[a]*
1	2.04	122	1022	0.12
2	5.31	268	465	0.58
3	9.12	456	139	3.28
4	12.7	635	648	0.98
5	9.07	453	1303	0.35
6	2.31	116	437	0.27

[a] The labeled rate is 2 lbs of granular material per 1000 ft^2 of turf providing 9.7 g/m^2. The theoretical mass of the active ingredient (diazinon) per square meter of turf, based on labeled rates is 0.5 g/m^2.

Concentrations Measured from Soil Cores

Some homes exhibited background concentrations of diazinon in soil prior to the application (Table 5). Following the granular applications the concentrations increased sharply above background, peaked by day 2 for the majority of homes and declined at rates ranging from 25 to 96% by day eight. Despite the temporal decline the measured concentrations remained high eight days following the application relative to the measured background concentrations. The soil measurements identify a source of diazinon from which residues might emanate. These measurements are in line with the reported half life of diazinon in soil between 7 to 52 days depending on environmental conditions such as soil type, temperature and moisture content *(29, 30)*. The 1-cm deep soil cores taken in this study provide a reasonable estimate of the of the total mass present in the soil column, as Kuhr and Tashiro *(29)* reported that very little diazinon could be measured deeper than 1.3 cm below the soil surface.

Table 5. Diazinon measured from soil (ng/g) following granular applications to residential lawns.

House	*Day*				
	Pre	*1*	*2*	*4*	*8*
1	<1.7	3800	2070	2150	1680
2	<1.7	19995	14757	12640	14213
3	6.1	72375	74954	55044	54230
4	<1.7	38213	64037	13744	24409
5	183	33230	27986	9200	1070
6	8.3	45015	41753	98968	16078
Avg.		35438	37592	31957	18613

As expected, the total mass applied influenced the measured soil concentrations. A significant association was shown between the (log transformed) post-application soil concentrations and the (log transformed) mass applied per unit area (ANOVA, $r^2 = 0.43$, $p = 0.0003$). The size of the treated area itself was also a strong indicator of soil concentrations, with the larger the areas associated with lower soil concentrations ($p = 0.0006$). Factors such as the applicator's technique and the spatial distribution of the granules within a confined area might be important elements along with housing factors in future observational studies of potential human exposure.

Turf Transferable Residues

Turf transferable residues concentrations are shown in Table 6. With the exception of one house, diazinon levels prior to applications were typically below the detection limit (0.003 ng/cm^2). Applications at all homes resulted in an increase in measurable diazinon transferred from turf to the PUF roller. The highest turf transferable residues were typically measured between 1 to 2 days post application and then declined over the remainder of the study. However, in most cases, transferable residues remained up to two orders of magnitude higher than pre-application levels through 8 days. Transferable residue levels declined more rapidly than soil concentrations (Figure 1). The relatively low transferable residue levels observed agree with previous work by Sears *(31)* who found that despite the abundant mass of diazinon residue present in the thatch following an application to turf, only 10% could be dislodged using vigorous wiping techniques. Similarly, residues declined rapidly within the first day of application and slowly declined through 15 days post application.

Table 6. Turf transferable residues measured using the PUF roller following applications to lawns.

	Day (ng/m^2)				
House	*Pre*	*1*	*2*	*4*	*8*
1	0.19	22.1	51.2	0.43	1.9
2	<0.003	48.6	7.4	10.6	1.7
3	<0.003	NC[a]	970.1	26.7	66.3
4	<0.003	75.1	7.1	2.9	0.5
5	<0.003	0.2	0.2	0.1	0.004
6	<0.003	9.1	6.3	7.2	3.6

[a] NC denote that the sample was not collected.

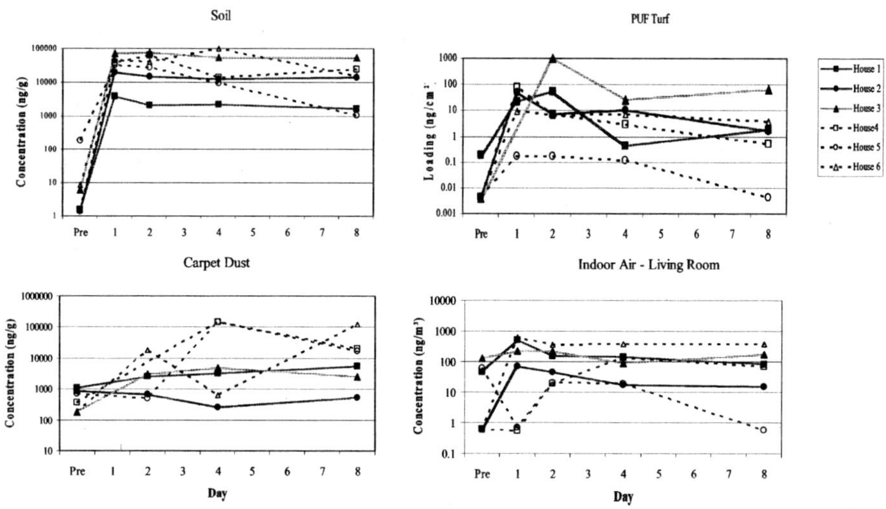

Figure 1. A comparison of temporal changes in diazinon levels from indoor air, carpet dust, soil and transferable residues following an application of a granular formulation to residential lawn.

Doormat Sweepings

The doormat was deployed as a simple tool to demonstrate the movement of residues from the source to the principal route of ingress and egress from the residence by the occupants and their pet dogs (Table 7). While doormat levels generally declined between days 2 and 8, the levels measured in the doormat sweepings at one home were actually 32% higher at day 8 than those measured at day 2. This increase did not correspond to any increase in soil concentration or transferable residue levels. In this case, the doormat loading likely was a function of occupant and pet activity during that time. Although no apparent relationship was observed between outdoor soil concentrations and the measured doormat loadings, the levels measured on the doormat provided evidence that residues were deposited at or transferred to the entryway of every single home studied.

Table 7. Diazinon measured from particles vacuumed from door mats (doormat sweepings) of homes following turf applications.

	Day (ng/m^2)		
House	*2*	*4*	*8*
1	69934	22274	14353
2	482	1079	719
3	213288	133484	21184
4	720007	442972	72376
5	1711656	220933	410559
6	44574	230337	35356

Indoor Air

Three homes had measurable background levels of diazinon prior to the lawn application. As was the case for residues in other media, the indoor air concentrations typically declined after day 2. Maximal concentrations were measured at days 1 and 2. The airborne levels were statistically different between all homes ($p < 0.0001$). A statistically significant difference was observed between the living room and bedroom ($p < 0.05$), suggesting that the level of occupancy, by both pets and participants, may affect the concentration in a particular room within the home following an outdoor application (Figure 2). Indoor concentrations were significantly ($p < 0.05$) and positively associated with indoor temperature. These findings suggest that seasonal temperature variations both indoor and outdoor might influence residential exposure to volatile or semivolatile insecticides.

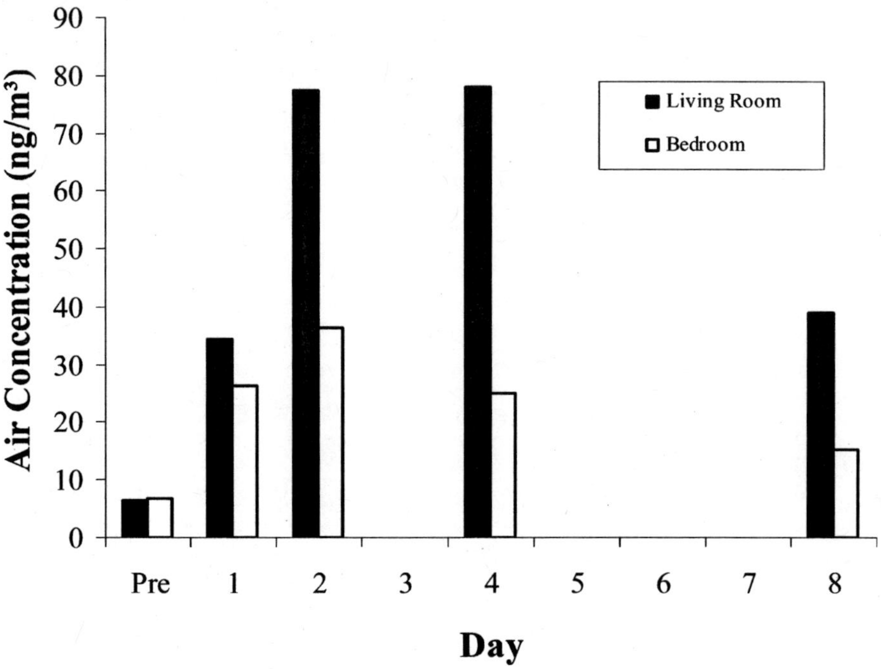

Figure 2. A comparison of the geometric mean of indoor air concentrations measured from the living rooms and bedrooms of six homes following applications of diazinon to residential turf.

Indoor Transferable Residues

Indoor PUF roller measurements exhibited a low detection frequency (27%) and are not presented. All measurements at two of the homes were below the limit of detection (0.006 ng/cm^2). The highest value (1.4 ng/cm^2) was observed at the home with the overall highest soil and turf transferable values (house 3).

The PUF roller has been shown to have a low collection (transfer) efficiency of less than 2% for both turf and indoor carpeting *(10, 32, 33,)*. The combination of low indoor surface loadings and low overall efficiency of the method makes the PUF roller a poor approach for estimating (indoor) transferable residues in an outdoor application scenario. Since the value is a common metric in residential exposure models the utility of the methods would benefit by enhanced collection efficiency.

Vacuum Dislodgeable Particles (Carpet Dust)

The elevated background concentrations of diazinon residues in household carpet dust (Figure 1) may be attributable to previous applications, as the persistence of some organophosphate pesticides in the indoor environments is well documented *(6, 11)*. The highest residue levels were typically measured at 2 to 4 days following the application. Applications to the turf contributed to the diazinon measured in the homes; however, the rate and magnitude of the increase relative to pre-application levels varied among homes. The variability may be associated with housing factors, the sampling location within rooms, or the occupants' movements or activities within, into, and around the home. Once the contaminant enters the home, indoor dust becomes a sink for diazinon and subsequently serves as a potential source for human exposure.

Media Comparisons

Comparisons among the concentration profiles in the different media suggest a movement of residues from the treated turf to the indoors. Figure 1 illustrates the residue levels from the treated soil, turf transferable (or PUF roller), indoor air and carpet dust measurements. All metrics exhibited an increase in measured concentration above background. Although soil concentration declined gradually over the eight day sampling period, the soil remained a continuous source of residues throughout the study. Transferable turf measurements showed a more rapid decrease beyond two days following the application. Diazinon, which is moderately water soluble, may be expected to migrate into the soil column to a depth of nearly 1.3 cm *(29)* as it is generally watered in following turf applications. Although the soil measurements are reflective of the source strength, the transferable residue levels suggest that the residues become less available for transfer beyond two days, perhaps due to leaching into the soil column. As discussed previously, residues measured from doormats suggest migration to the indoor/outdoor interface of the homes. Based on the vapor pressure of diazinon (2×10^1 mPa at 25 °C) *(34)*, volatilization from the application surface into the air, as well as subsequent infiltration of some fraction of the total volatilized mass is expected. In this study, concentrations of diazinon in indoor air increased rapidly (within 24-h following the application) and remained elevated over the 9-day study period. In contrast, carpet dust concentrations generally did not reach maximal concentrations until

2 to 4 days post application. Relative to air, carpet dust displayed a slower rate of intrusion that is probably dependent on tracked-in particles linked to the outdoor activities of the occupants. Additional analysis of the findings shows that doormat loadings were a highly significant predictor of indoor carpet dust ($p < 0.0001$) suggesting a strong relationship between doormat particle deposition and indoor dust loadings in the rooms most commonly occupied by participants and their pet dogs. Soil concentrations, on the other hand, trended towards the prediction of indoor dust loadings but the relationship was not statistically significant.

Exposure and Absorbed Mass Estimates for Children

Figure 3 represents the percentage of estimated absorbed mass apportioned by exposure route (excluding dietary ingestion) before and after the application. Before application when background concentrations are low, the primary route of exposure to diazinon was likely inhalation with an estimated of 49% of total accountable absorbed mass, while hand-to-mouth ingestion potentially accounted for 33% and dermal about 18%. However, following the application hand-to-mouth ingestion is estimated to have increased sharply to 63% of the total, dermal exposure also increased to 35% while inhalation exposure declined to 2%. Examination of the per cent contribution by day (Figure 4) further suggests that the contribution through inhalation becomes nearly negligible immediately after the application. Furthermore, the potential contribution through dermal exposure reached its maximal levels by two days post application followed by a decrease. Potential exposure via the gastrointestinal route remained important over the duration of the study. The relatively large potential intake through non-dietary ingestion is a function of the soil and carpet dust concentration and the transfer coefficient. In light of the attribution of the majority of adsorbed mass to hand-to-mouth ingestion, it should be noted that the values employed for the transfer coefficients are highly uncertain.

The calculated absorbed mass for each participant at each sampling interval is illustrated in Figure 5. Considerable variation can be observed among the six participants. In addition to the amount of diazinon applied and measured diazinon levels, these differences may result from individual activities on the treated turf, the transport of residues into homes and highly variable housing factors. Generally the absorbed mass estimates coincide with the maximal concentrations measured from the turf (both soil and turf transferable).

The total amount of diazinon absorbed from all exposure routes excluding dietary on days following application ranged from 0.271 to 173 ng/day, with a median value of 17.8 ng/day. Since dietary ingestion has been previously reported as representing the dominant route of exposure to diazinon in the absence of any recent application *(35)*, a comparison to dietary exposure is warranted. No dietary samples were collected in this study, and published diazinon ingestion data are sparse. However, Morgan and colleagues *(36)* estimated a median dietary intake of approximately 12 ng/day (maximum = 61 ng/day) among 111 children in North Carolina and Ohio enrolled in the CTEPP Study *(36)*, and Moschandreas and colleagues *(37)* estimated a median dietary exposure of 0.55 ng/kg/day for Arizona children and adults, corresponding to a range of 9.4 to 20 ng/day for the individuals in this study. The estimated daily amount of diazinon absorbed into the body as a direct result of outdoor turf applications was similar to the amount typically absorbed through dietary ingestion. For the most highly exposed participant, however, the potential contribution due to turf treatment was approximately eight times that of the estimated contribution due to dietary ingestion on days 2 though 8.

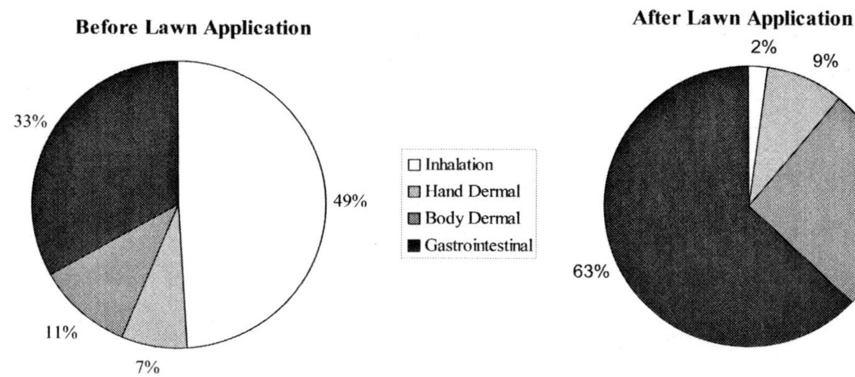

Figure 3. The principal routes of the children's exposure to diazinon prior to and following a granular application to residential lawns.

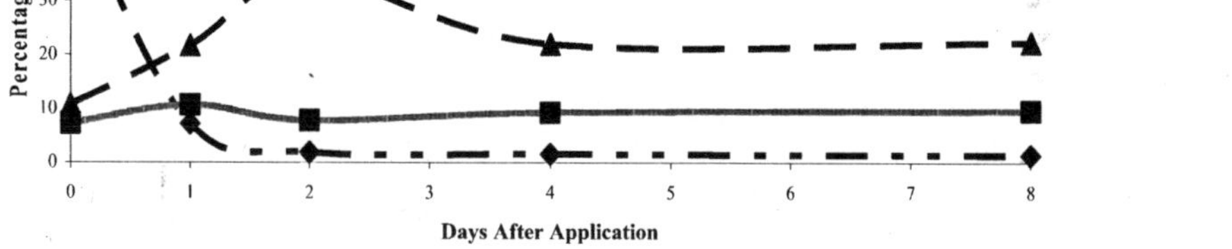

Figure 4. The changes in the contribution of each exposure pathway over time following a granular diazinon application to residential lawns.

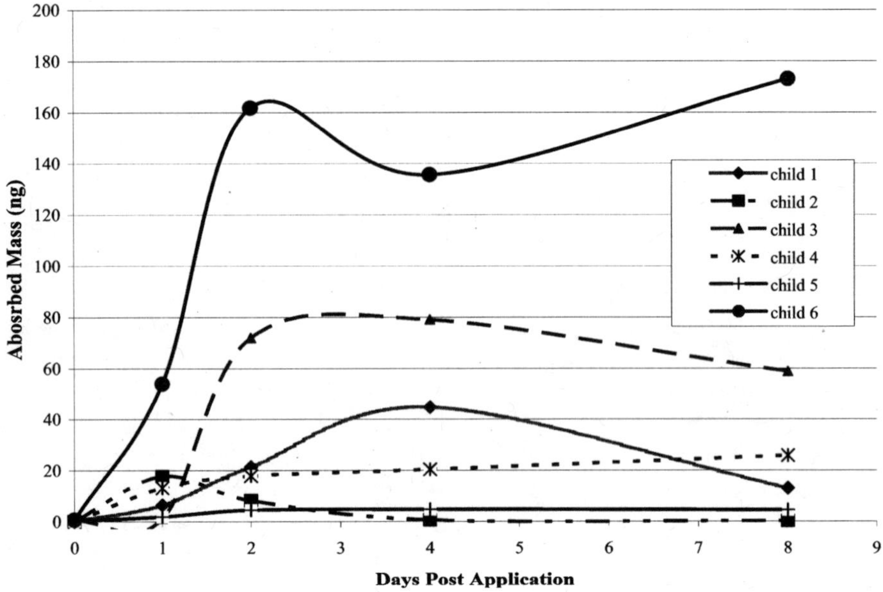

Figure 5. The estimated absorbed mass of diazinon by day for each participant following a granular application to residential lawns.

Summary

Insecticide applications to lawns, although uncommon in many countries, are very common in the U.S. This study examined turf applications of granular diazinon, a semivolatile organophosphate insecticide, which has since been deregistered for use on residential lawns. We can derive only rudimentary inferences from these findings due to the small sample size. We anticipated that semivolatile compounds are transported from the source into the indoor living area and that exposure by inhalation would be an important route of exposure. Our results, based on modeling, suggest that the insecticide residues do indeed infiltrate the home, but that inhalation is only a minor potential route of exposure after outdoor lawn applications.

The concentration-time profiles in the various exposure media (Figure 1), particularly for carpet dust suggest that for this type of application, track-in may be the most important mechanism for translocation. The early peak of indoor air concentrations relative to dust concentrations provides evidence of an intrusion of volatilized diazinon.

The utility of the doormat sweeping in establishing a relationship between outdoor and indoor concentrations is equivocal, as the doormat levels are highly predictive of carpet dust concentrations but not predictive of indoor transferable residues levels. Nonetheless the doormats provided supportive evidence that pesticide residues were deposited at least to the outdoor/indoor interface and potentially migrated further into the homes.

SHEDS model estimates suggest that home occupants might experience exposure to this pesticide following outdoor applications. While this exposure may be primarily through non-dietary (i.e. hand-to-mouth) ingestion, the dermal route, and (to a lesser extent) the inhalation route also contribute to the total absorbed mass. The estimated daily absorbed mass declined little over the nine days of the study. It is unlikely that the typical consumer of such products realizes the extent to which the homes become contaminated and the occupants exposed following outdoor applications.

Acknowledgements

Gratitude is extended to the Carry Croghan, April Coan and Don Whitaker of the US EPA who aided in the success of this study. Appreciation is extended to Bill Ellenson, Chris Fortune, Zora Drake and Fred Blanchard of ALION (formerly ManTech) for carrying out participant recruitment and the field component in its entirety. In addition, we would like to thank Bo Wang and Tom McDonald of TDI-Brooks for carrying out the chemical analysis of all samples.

Disclaimer: The United States Environmental Protection Agency through its Office of Research and Development funded and managed the research described here. It has been subjected to Agency's administrative review and approved for publication.

References

1. Kiely, T.; Donaldson, D.; Grube, A. Pesticides Industry Sales and Usage, 2000-2001 Pesticides Market Estimates, U.S. Environmental Protection Agency, Office of Pesticide Programs, Washington, DC, 2004.
2. U.S. Environmental Protection Agency. 1985. Diazinon Pesticide fact sheet 96.1. USEPA, Office of Pesticide Programs, Washington DC.
3. U.S. Environmental Protection Agency: http://www.epa.gov/oppsrrd1/REDs/diazinon_ired.pdf. March 27, 2006. Edwards, D., Finalization of Interim Reregistration Eligibility Decisions (IREDs) and Interim Tolerance Reassessment and Risk Management Decisions (TREDs) for the Organophosphate Pesticides, and Completion of the Tolerance Reassessment and Reregistration Eligibility Process for the Organophosphate Pesticides, Memorandum to Jim Jones Dated July 31, 2006, Environmental Protection Agency, office of Prevention, Pesticides and Toxic Substances, Washington, DC.
4. Wright, C. G.; Leidy, R. B. Chlorpyrifos residues in air after the application to crevices in rooms. *Bull. Environ. Contam. Toxicol.* **1978**, *20*, 340-344.
5. Leidy, R. B.; Wright, C. G.; Dupree Jr., H. E. Exposure levels of to indoor pesticides. In *Pesticides in Urban Environments*; Racke, K. D., Leslie, A.

R., Eds.; ACS Symposium Series 522. American Chemical Society, Washington DC, 1993; pp. 282-295.

6. Gurunathan, S.; Robson, M.; Freeman, N.; Buckley, B.; Roy, A.; Meyer, R.; Bukowski, J.; Lioy, P. J. Accumulation of chlorpyrifos on residential surfaces and toys accessible to children. *Environ. Health Perspect.* **1998**, *106* , 9-16.
7. Lewis, R. G.; Fortune, C. F.; Blanchard, F. T.; Camann, D. E. Movement and deposition of two organophosphate pesticides within a residence after interior and exterior applications. *J. of Air Waste Manag. Assoc.* **2001**, *51*, 339-351.
8. Hore, P. Pesticide accumulation patterns for child accessible surfaces and objects and urinary metabolite excretion by children for two weeks after a professional crack and crevice application. Ph.D. Thesis, Rutgers University, New Brunswick, NJ. 2003.
9. Stout II, D. M.; Mason, M. A. The distribution of chlorpyrifos following a crack and crevice type application in the US EPA indoor air quality research house. *Atmos. Sci.* **2003**, *27*, 5539-5549.
10. Lewis, R.G., 2005. Residential Post Application Exposure Monitoring, Chapter 3 in C.A. Franklin and J. P. Worgan, eds., *Occupational and Incidental Residential Exposure Assessment*, John Wiley & Sons, LTD.
11. Lewis, R. G.; Fortune, C. R.; Camann, D. E. Evaluation methods for the monitoring of the potential exposure of small children to pesticides in the residential environment, *Arch. Environ. Health Perspect.* **1994**, *26*, 37-46.
12. Whitmore, R. W.; Immerman, F. W.; Camann, D. E.; Bond, A. E.; Lewis, R. G.; Schaum, J. L., Non-occupational exposures to pesticides for residences of two U.S. cities. *Arch. Environ. Contam. Toxicol.* **1994**, *26*, 47-59.
13. Lewis, R. G.; Macleod, K. E. Portable sampler for pesticides and semivolatile industrial organic chemicals in air, *Anal. Chem.* **1982**, *54*, 310-315.
14. Stout II, D. M.; Leidy, R. B. A preliminary examination of the translocation of microencapsulated cyfluthrin following applications to the perimeter of residential dwellings. *J. Environ. Sci. Health, Part B*, **2000**, *B35* , 477-489.
15. Nishioka, M. G.; Burkholder, H. M.; Brinkman, M. C.; Lewis R. G. Distribution of 2,4-dichlorophenoxyacetic acid in floor dust throughout homes following homeowner and commercial lawn applications: quantitative effects on children, pets and shoes. *Environ. Sci. Tech.* 1996, 30, 3313-3320.
16. Nishioka, M. G.; Burkholder, H. M.; Brinkman, M. C.; Gordon, S. M.; Lewis R. G. *Simulation of track-in of lawn applied pesticides from turf to home: comparisons of dislodegable turf residues from carpet dust and carpet surface residues*. EPA/600/R-97/108, United States Environmental Protection Agency, National Exposure Research Laboratory, Research Triangle Park, NC, 1997.
17. Nishioka, M. G.; Lewis R. G.; Brinkman, M. C.; Burkholder, H. M.; Hines C. E.; Menkedick, J. R. Distribution of 2,4-D in air and on surfaces inside residences after lawn applications: comparing exposure estimates from

various media for young children. *Environ. Health Perspect.* **2001**, *109* , 1185-1191.

18. Nishioka, M. G.; Lewis R. G.; Brinkman, M. C.; Burkholder, H. M. Foot transfer of lawn-applied pesticides from turf to carpet: comparison of semivolatile chlorpyrifos with non-volatile chlorothalonil. *Bull. Environ. Contam. Toxicol.* **2002**, *68*, 64-71.
19. Morgan, M. K.; Stout II, D. M.; Wilson, N. K.; Feasibility study of the potential for human exposure to pet-borne diazinon residues following lawn applications. *Bull Environ. Contam. Toxicol.* **2001**, *66*, 295-300.
20. ASTM 2004-. Standard Practice for the Collection of Dislodgeable Rresidues from Floors. D 6333-98 (Rev 2004), In *Annual Book of ASTM Standards*, Vol. 11.03: ASTM International, West Conshohoken, PA; 2007.
21. ASTM. Standard Practice for the Sampling and Selection of Analytical Techniques for Pesticides and Polychlorinated Biphenyls in Air. D4861-06, *Annual Book of ASTM Standards*, Vol. 11.03: ASTM International, West Conshohoken, PA, 2007.
22. ASTM. Standard Practice for the Collection of Dust from Carpeted Floors for Chemical Analysis. D5438-06. In *Annual Book of ASTM Standards*, Vol. 11.03: ASTM International; West Conshohoken, PA: 2007.

23. Glen, G. (2007). SHEDS-Multimedia version 3 SAS code.

24. Zartarian, V., Xue, J., Ozkaynak, H.; Dang, W., Glen, G., Smith, L., Stallings, C. A probabilistic exposure assessment for children who contact CCA-treated playsets and decks, Using the Stochastic human Exposure and Dose Simulation Model for the Wood Preservation Exposure Scenario (SHEDS-WOOD), Draft preliminary report, Sept. 25, 2003.

25. McCurdy, T.; Glen, G.; Smith, L.; Lakkadi, Y. The National Exposure research Laboratory's consolidated human activity database. *Journal of Exposure Anal. Environ. Epidemiol.* **2000**, *10*, 566-578.
26. Zartarian, V., Xue, J., Ozkaynak, H. 2002. Quantifying aggregate chlorpyrifos exposure and dose to children using a physically based two-stage Monte Carlo probabilistic model. Society for Risk Analysis Conference, New Orleans, LA, 8-11 December.
27. Xue, J.; Zartarian, V.; Özkaynak, H.; Dang, W.; Glen, G.; Smith, L.; Stallings, C. A probabilistic arsenic exposure assessment for children who contact chromated copper arsenate (CCA)-treated playsets and decks: Part 2: Sensitivity and uncertainty analyses. *Risk Analysis* **2006**, *26*, 533-541.

28. Zartarian, V.; Ozkaynak, H.; Dang, W.; Graham G.; Smith, L.; Stallings, C. A probabilistic arsenic exposure assessment for children who contact CCA-treated playsets and decks, part 1: model methodology, variability results, and model evaluation. *Risk Anal.* **2006**, *26*, 515-531.
29. Kuhr, R. J.; Tashiro, H. Distribution and persistence of chlorpyrifos and diazinon applied to turf. *Bull. Environ. Contam. Toxicol.* **1978**, *20*, 652-656.
30. Michel Jr., F. C.; Reddy, C. A.; Forney, L. J. Fate of carbon-14 diazinon during the composting of yard trimmings. *J. Environ. Qual.* **1997**, *26*, 200-205.

31. Sears M. K.; Bowhey, C.; Braun, H.; Stephenson, G. R.; Dislodgeable residues and persistence of Diazinon, chlorpyrifos and isofenphos following their application to Turfgrass. *Pestic. Sci.* **1987**, *20*, 233-231.
32. Fortune, C. R. *Evaluation of Methods for Collecting Dislodgeable Pesticide Residues from Turf.* EPA/600/SR-97/119, United States Environmental Protection Agency, National Exposure Research Laboratory: Research Triangle Park, NC, 1998.
33. Fortune, C. R. *Round Robin Testing of Methods for Collecting Dislodgeable Residues from Carpets.* EPA/600/SR-97/107, United States Environmental Protection Agency, National Exposure Research Laboratory: Research Triangle Park, NC, 1998.
34. Tomlin C. D. S. editor. 2006. *The Pesticide Manual, 14th Ed.*; Alton, Hampshire, UK: British Crop Protection Council, Surrey, pp. 264-265.
35. Morgan M .K.; Sheldon L. S.; Croghan C. W.; Chuang, J. C.; Lordo, R. A.; Wilson, N. K.; Lyu. C.; Brinkman, M.; Morse, N.; Chou, Y. L.; Hamilton, C.; Finegold, J. K.; Hand, K.; Gordon, S. M. (2004) A Pilot Study of Children's Total Exposure to Persistent Pesticides and Other Persistent Organic Pollutants (CTEPP). U.S. Environmental Protection Agency, Office of Research and Development, Washington, DC. EPA/600/R-041/193.
36. Morgan M .K.; Jones, P. A.; Egeghy, P. P.; Croghan, C. W.; Sheldon L. S.; Barr, D. B. (2007) Cumulative exposures of 111 preschool children to chlorpyrifos and diazinon in their everyday environments. Presented at the International Society of Exposure Analysis conference, Durham, NC. October 14, 2007.
37. Moschandreas, D. J.; Kim, Y.; Karuchit, S.; Ari, H.; Lebowitz, M. D.; O'Rourke, M. K.; Gordon, S.; Robertson, G. In-residence, multiple route exposures to chlorpyrifos and diazinon estimated by indirect method models. *Atmos. Environ.* **2001**, *35,* 2201-2213.
38. Nolan, R. J.; Rick, D. L.; Freshour, N. L.; Saunders, J. H. Chlorpyrifos: pharmacokinetics in human volunteers. *Toxicol. Appl. Pharmacol.* **1984**, *73*, 8-15.
39. Wester, R. C.; Sedik, L.; Melendres, J.; Logan, F. Maibach, H. I.; Russell, I. Percutaneous absorption of diazinon in humans. *Food Chem Toxicol.* **1993**, *31*, 569-572.
40. Poet, T. S.; McDougal, J. N. Skin absorption and human risk assessment. *Chem.-Biol. Interact.* **2002**, *140*, 19-34.
41. Poet, T. S.; Kousba, A. A.; Dennison, S. L.; Timchalk, C. Physiologically based pharmacokinetic/pharmacodynamic model for the organophosphorus pesticide diazinon. *Neuro. Toxicol.* **2004**, *25*, 1013-1030.
42. Ozkaynak, H.; Xue, J.; Zartarian, V.; Graham, G.; Luther, S. Presented at 15th Annual ISEA Conference, Tucson Arizona, October 30-November 3, 2005.

Chapter 12

Exposure of Adults and Children to Organophosphorus Insecticides used in Flea Collars on Pet Dogs

Janice E. Chambers[1] and M. Keith Davis[1]

[1]Department of Basic Sciences/Center for Environmental Health Sciences, College of Veterinary Medicine, Mississippi State University, Mississippi State, MS 39762

Our laboratories have investigated the exposure of people to the organophosphorus insecticides chlorpyrifos and tetrachlorvinphos that were contained in flea collars used on their pet dogs. Long-term studies conducted over much of the recommended useful lifetime of the collars indicated that residues of both insecticides were transferable to white cotton gloves which were used to rub the fur of the dogs, with the tetrachlorvinphos residues considerably higher than the chlorpyrifos residues. In short-term studies, residues of both insecticides were transferred to tee shirts worn by the children. In these same studies the urinary metabolite of chlorpyrifos was perhaps slightly elevated over the background level in children but was not elevated over the background level in adults. In contrast the urinary metabolite of tetrachlorvinphos was substantially elevated over background levels in both children and adults. However, the significance of these findings to risk assessment is not known.

A large proportion of homes in the United States have pet dogs, roughly about 40%, and of those homes, about two thirds have children (*1*). Since ectoparasites are both a medical problem and a nuisance to pets and their owners, many pet owners use flea and tick control products. One type of flea and tick control product that is common, easy to obtain and available without a prescription (and therefore is relatively inexpensive) is a plastic collar embedded with an insecticide.

Organophosphorus insecticides have been the insecticides of choice for a number of these collars. Organophosphorus insecticides or their active metabolites (i.e., oxons) are anticholinesterases; the insecticides or their oxons phosphorylate the active site serine of the acetylcholinesterase, a reaction that inhibits the catalytic action of the enzyme allowing the accumulation of the neurotransmitter acetylcholine in synapses and neuromuscular junctions (*2*). At sufficiently high levels of acetylcholinesterase inhibition, a variety of autonomic nervous system effects occur as can tremors and convulsions, with death occurring from respiratory failure at lethal levels of exposure. The organophosphorus insecticides have been subject to the Food Quality Protection Act-mandated cumulative risk assessment as these insecticides act toxicologically through a common mechanism of toxicity, i.e., the inhibition of acetylcholinesterase. As routes of exposure are considered for a cumulative risk assessment, certainly any sources of insecticide from residential uses would need to be considered. The use of insecticides for flea and tick control on pets would be a potential source of insecticide exposure in residential settings, and this potential exposure has formed the basis for the study of several flea and tick control products in our laboratories.

Little is known about how much insecticide might transfer to people in contact with their pets from any topically-applied insecticide (such as from dips or shampoos) or any insecticide from a product in which the insecticide diffuses out of the product and deposits on the fur (such as from collars). In the case of some of these products, the concentration of insecticide is quite high either directly deposited on the animal's fur or embedded in the product, and could provide a source of appreciable exposure to the people in contact with their pets. Information on this potential source of exposure is necessary in order to determine whether this exposure is of a magnitude of concern, and, if so, for this source to be included in the cumulative risk assessments.

Protocols for the standard measurement of transferable residues from such ectoparasite treatments were not established when our group started these studies, so we developed protocols involving petting the treated dogs with new laundered and solvent-extracted white cotton gloves for a set 5-minute period in a prescribed area of the dog's fur. The gloves were subsequently extracted with appropriate solvents, and the residues were quantified by analytical chemistry methods. These data were considered to represent an estimate of the level of residues that were available for transfer from the fur of the dog to a person. Our initial studies were on two over-the-counter dips containing either chlorpyrifos or phosmet (*3*, *4*). Subsequently we studied two collars that contained either chlorpyrifos or tetrachlorvinphos (*5*, *6*). Concurrently with these latter two studies of collars, we were also able to do biomonitoring of urinary metabolites of the insecticide in an adult and in a child in each household participating in the study. For all of these studies the Institutional Animal Care and Use Committee (IACUC) of Mississippi State University provided prior approval of the protocols used for the handling of the dogs, and the Institutional Review Board (IRB) for Research involving Human Subjects of Mississippi State University provided prior approval of the protocols used for the recruitment and tests involving human subjects. The informed consent of the adult participants and the assent of the children were obtained prior to their participation.

The protocol we developed to quantify transferable residues from the dog's fur was to rub the dog in a back-and-forth manner for a 5-minute period in a set measured area of the dog's fur. These gloves were light-weight white cotton gloves, not previously used, which were laundered and solvent-extracted prior to use. The sampling occurred at three locations on the dog: on a place on the back near the tail, distant from the collar; on the neck with the collar removed; and on the neck with the collar in place (over the collar). The gloves were then placed into glass bottles which had been pre-rinsed with solvent. Standard solvent extractions occurred, and the insecticide was quantified using gas chromatography with electron capture detection. (We wish to note that the dogs seemed to be very willing participants for these sampling sessions).

For the studies where we conducted biomonitoring, in the participating household an adult (either sex) and a child (either sex) between the ages of 3 and 12 years provided first morning void urine samples. These samples were acid-hydrolyzed and extracted, and the insecticide metabolite was quantified using gas chromatography with electron capture detection (chlorpyrifos metabolite) or gas chromatography with mass spectrometry detection (tetrachlorvinphos metabolite). In these studies the participating child also wore a new laundered white cotton tee shirt for several hours on the afternoon prior to the collection of the urine sample. A section was cut from the front of the tee shirt and, similar to the procedure with the gloves, the fabric was extracted with appropriate solvents, and the insecticide was quantified using gas chromatography with electron capture detection. The rationale for the tee shirt studies was that the tee shirt might be a suitable surrogate for the potential transferable residues and therefore for the level of exposure. Statistical analysis of the data tested for correlations between the residues observed and certain characteristics of the dog such as fur length or behavioral characteristics such as the length of time the child spent with the pet dog.

Chlorpyrifos

Chlorpyrifos is a very widely used insecticide, and it displays a moderate level of acute oral mammalian toxicity but a low level of acute dermal toxicity (rabbit dermal LD_{50} for chlorpyrifos of 2 g/Kg) (*7*). While its uses have been curtailed in recent years, it has been so widely used in the past that most of the United States population seems to have exposure to chlorpyrifos, as evidenced by the presence of the chlorpyrifos metabolite trichloropyridinol (TCP) in the majority of the American population sampled in such biomonitoring studies as the National Health and Nutrition Examination Survey (NHANES) (*8*).

We performed two studies using an over-the-counter chlorpyrifos-containing collar which contained 8% chlorpyrifos (*5*). The long-term study had sampling of transferable residues for 168 days, which was a substantial fraction of time of the 11 months for which the collar was recommended. The short-term study selected a week from the data of the long term study (week 3 after collar placement) during which peak transferable residues occurred. Transferable residues from the dog's fur were obtained on two occasions during this week. Urine samples were obtained from an adult and a child in the household of the dog on 5 consecutive days with the child wearing the white cotton tee shirts on 5 consecutive days using the days preceding the urinary metabolite collection.

Pretreatment residues on gloves (*i.e.*, background levels) for the five minute rubbing period were measurable, but low, between 1 and 2 μg/glove. Transferable residues increased over the first few days following collar placement, and then remained relatively constant over the rest of the 168 day sampling period. As expected, transferable residues obtained directly over the collar were the highest (about 170 to almost 400 μg/glove), followed by residues from the neck without the collar in place (about 50 to 250 μg/glove), and lowest from the back distant from the collar (from 2 to 14 μg/glove (Table I). The data indicated that relatively low levels of insecticide migrated into the fur a long distance from the collar. The same patterns were apparent in the short term study which obtained transferable residues from fur on days 14 and 20 following collar placement: neck with collar in place, 430-500 μg/glove; neck with collar removed, 280-350 μg/glove; and back, 7-10 μg/glove (Table II).

In the short-term study, about 120-200 ng chlorpyrifos/g shirt was observed (Table III). It was originally thought that the tee shirt might be a useful surrogate for biomonitoring data, since obtaining urine samples is somewhat problematical in children; however, the data from the tee shirts did not correlate well with the urinary metabolite data or with any of the activity records provided by the parents on length of time the child spent with the pet dog.

Table I. Chlorpyrifos Concentrations on Cotton Gloves for the Long-term Study (concentration ± standard error)

Day	*Back (μg/glove)*	*Neck (μg/glove)*	*Collar (μg/glove)*
0 (4 Hour)	2.17 ± 1.01	49 ± 8	168 ± 27
1	4.86 ± 1.90	128 ± 35	190 ± 29
3	5.73 ± 1.00	125 ± 19	218 ± 25
7	4.11 ± 0.59	118 ± 15	184 ± 21
14	6.30 ± 1.51	242 ± 34	391 ± 75
28	12.65 ± 3.81	241 ± 38	318 ± 27
56	8.45 ± 1.98	238 ± 41	350 ± 44
84	14.18 ± 3.60	216 ± 26	310 ± 40
112	10.15 ± 1.77	252 ± 38	387 ± 49
140	10.01 ± 2.43	220 ± 26	377 ± 61
168	9.16 ± 2.50	194 ± 22	313 ± 34

NOTE: Pretreatment value was 1.74 ± 0.49 μg/glove.

Table II. Chlorpyrifos Residues on Cotton Gloves for the Short-term Studies (concentration ± standard error)

	Day	*Back (μg/glove)*	*Neck (μg/glove)*	*Collar (μg/glove)*
Chlorpyrifos	14	10.14 ± 2.31	356 ± 66	503 ± 86
	20	6.53 ± 1.73	279 ± 60	434 ± 80

NOTE: Pretreatment value for chlorpyrifos was 1.43 ± 0.77 μg/glove.

Table III. Chlorpyrifos Tee Shirt Residues for the Short-term Studies (concentration ± standard error)

	Day	*ng/g Shirt*
Chlorpyrifos	15	134.06 ± 66.03
	16	201.46 ± 67.15
	17	126.12 ± 25.81
	18	133.89 ± 45.49
	19	172.03 ± 68.96

Even though chlorpyrifos uses have been reduced in recent years, it is still a widely used insecticide and many, if not most, people have exposure to it, and its metabolite, TCP, is prevalent in human urine. For this reason, urine samples taken from the subjects, both adults and children, had non-zero levels of TCP, with averages across both the long term and short term studies of 8.0-9.2 ng/ml in adults and 10.5-13.5 ng/ml in children (Tables IV and V). In the long term (168 day) study, urinary TCP levels averaged 8.7-10.9 ng/ml for adults and 11.7-16.0 in children, with samples taken on 5 occasions during the 168 day test (Table IV). During the short term study, we concentrated on daily samples for 5 days to assess day-to-day variation within individuals; samples were taken during the third week following placement of the collar on the pet dog. Urinary TCP levels were from 6.9-9.9 ng/ml in adults and from 11.2-15.9 ng/ml in children (Table V). In both of these studies the levels of TCP were higher in the urine from children than in the urine from adults. None of the post-treatment values were significantly different ($P < 0.05$) from the pretreatment values in either the long term or the short term studies in either adults or children. In both studies, the post-treatment adult TCP levels were both above and below the pretreatment levels, so it appears that no enhanced exposure of adults occurred from the flea collar on the pet dog. However, in the long term study, the post-treatment children's urinary TCP levels were generally higher than the pre-treatment value, even though not statistically significant, while in the short term study, the post-treatment values were both above and below the pretreatment values. Therefore, it is difficult to conclude whether there is enhanced exposure of children to chlorpyrifos from this flea collar because one study suggested there was, but the other study suggested there was not. Certainly it is logical to think that children would in all likelihood be in greater contact with a pet dog than the adults of the household, so there is concern regarding this as a potential exposure route, but the numbers are not convincing that this is the case. The fact that the pretreatment TCP level in children in the long term study (10.5 ng/ml) was lower than the pretreatment value in children in the short term study (13.5 ng/ml) suggests the possibility that the apparent greater levels in children in the long term study are an artifact of the lower pretreatment number which was used for the comparison. The pretreatment level in children for the short term study (13.5 ng/ml) is in the middle of the range of the long term study post-treatment values. Therefore, we can conclude that adults do not seem to receive an enhanced exposure to chlorpyrifos from this collar. However, from these two studies we remain unable to conclude definitively whether children received an enhanced exposure to chlorpyrifos from this collar; however, if they did, it was only a very small amount and did not raise the level of exposure appreciably over the background levels.

Table IV. Urinary TCP Concentrations for the Long-term Study (concentration ± standard error)

	Day	*Adult*	*Child*
TCP (ng/mL Urine)	3	8.75 ± 1.35	11.70 ± 2.21
	7	9.43 ± 2.09	13.01 ± 2.04
	28	9.80 ± 1.42	16.01 ± 2.53
	84	10.88 ± 1.82	15.10 ± 1.95
	168	8.72 ± 1.34	12.08 ± 2.12

NOTE: Pretreatment values for TCP = 9.15 ± 1.62 for Adults and 10.49 ± 1.83 for children.

Table V. Urinary TCP Concentrations for the Short-term Studies (concentration ± standard error)

	Day	*Adult*	*Child*
TCP (ng/mL Urine)	16	6.92 ± 1.31	14.79 ± 2.88
	17	7.12 ± 1.49	13.57 ± 2.47
	18	9.05 ± 1.95	15.31 ± 2.39
	19	9.92 ± 2.18	15.92 ± 2.23
	20	9.68 ± 1.61	11.15 ± 1.18

NOTE: Pretreatment values for TCP = 7.95 ± 0.86 for adults and 13.54 ± 1.62 for children.

Tetrachlorvinphos

Another organophosphorus insecticide that has been used routinely in flea collars is tetrachlorvinphos. Tetrachlorvinphos (TCVP) is a low toxicity pesticide (rabbit dermal LD50 of 2.5g/Kg) (*9*). The TCVP collar was also an over-the-counter collar and contained 14% TCVP. It had a recommended use time of 4 months, so we conducted a long term study over 112 days (*6*). Similar to the protocol we described above, transferable residue samples were obtained by rubbing the dog's fur for 5 minute sampling periods over the same three regions of the dog: the back distant from the collar, the neck with the collar removed, and the neck with the collar in place. As expected, and similar to the results described above for chlorpyrifos, the lowest residues were obtained on the back distant from the collar and the highest residues were obtained by rubbing over the collar. In contrast to the chlorpyrifos collar, there was a peak of transferable residues obtained over the first two weeks after collar placement with considerably lower levels of residues after the first two weeks. Also in contrast to chlorpyrifos, the levels of transferable residues were considerably higher by about 2 orders of magnitude (Table VI).

Table VI. Tetrachlorvinphos Concentrations on Cotton Gloves for the Long Term Study (concentration ± standard error)

Day	*Back (μg/glove)*	*Neck (μg/glove)*	*Collar (μg/glove)*
0 (4 Hour)	185 ± 26	3,530 ± 564	14,340 ± 1,531
3	261 ± 52	8,042 ± 706	23,728 ± 2,125
7	177 ± 27	8,674 ± 860	24,039 ± 3,972
14	152 ± 22	6,062 ± 902	19,309 ± 3,252
28	144 ± 15	3,844 ± 597	12,568 ± 2,086
56	80 ± 18	2,802 ± 635	12,426 ± 2,362
84	36 ± 8	953 ± 168	4,956 ± 1,049
112	34 ± 8	549 ± 148	3,267 ± 982

NOTE: Pretreatment value was 0.22 ± 0.15 μg/glove.

A second study was also conducted with this collar, and, similar to above, was a short term study and collected tee shirt and biomonitoring data in addition to the transferable residues (*6*). This study was conducted over the second week after collar placement during the period that was identified in the long term study as the time of peak transferable residues. Transferable residues from the dog's fur were similar to those obtained in the long term study (Table VII). Tee shirt residues obtained from the front of the shirt which the child wore on the day before sampling for the urinary metabolite 2,4,5-trichloromandelic acid (TCMA) were in the range of about 1,000-2,000 ng/g shirt (Table VIII). These residues were about an order of magnitude greater than those obtained from the chlorpyrifos collar.

Table VII. Tetrachlorvinphos Residues on Cotton Gloves for the Short-term Studies (concentration ± standard error)

	Day	*Back (μg/glove)*	*Neck (μg/glove)*	*Collar (μg/glove)*
Tetrachlorvinphos	5	81.81 ± 19.0	9312 ± 1624	22,413 ± 2907
	12	82.12 ± 32.0	6738 ± 1091	15,788 ± 2101

NOTE: Pretreatment value for tetrachlorvinphos was 3.03 ± 1.53 μg/glove.

Table VIII. Tetrachlorvinphos Tee Shirt Residues for the Short-term Studies (concentration ± standard error)

	Day	*ng/g Shirt*
Tetrachlorvinphos	8	1,692 ± 657
	9	1,010 ± 435
	10	2,075 ± 1,031
	11	1,026 ± 277
	12	1,625 ± 926

The TCVP metabolite TCMA was quantified over 5 days in adults and children on the second week after placement of the collar. In contrast to chlorpyrifos, the pretreatment baseline levels of TCMA in both adults and children were very low, about 1.7 ng/ml (Table IX). These low baseline residues reflect the fact that TCVP is not a widely used insecticide and therefore the likelihood of TCVP exposure of people is very low. The residues of TCMA in adult urine were 43-104 ng/ml and the residues of TCMA in children's urine were 164-199 ng/ml in the post-treatment samples. These post-treatment samples are very clearly above the pretreatment values by one or two orders of magnitude, and were significantly different from pretreatment values ($P < 0.05$). Similar to the urinary metabolites from the chlorpyrifos collar, the TCMA levels in the urine of children were higher than those in adults. While the levels of these transferable residues and urinary metabolites seem very high, especially when compared to those from chlorpyrifos and TCP, TCVP is a very low toxicity insecticide and is probably metabolically detoxified very quickly, leading to the production of TCMA quickly. Therefore the significance of these high residues as related to hazard is unknown, and these numbers should not be construed at this point to be a cause for concern. If there are poor metabolizers in the population with respect to TCVP, these levels of residues might be of greater concern to such a sub-group, but we have no information about this possibility.

Table IX. Urinary TCMA Concentrations for the Short-term Studies (concentration ± standard error)

	Day	*Adult*	*Child*
TCMA (ng/mL Urine)	8	56.80 ± 13.91	175.04 ± 44.25
	9	65.74 ± 24.88	164.17 ± 43.91
	10	55.49 ± 18.88	172.57 ± 60.87
	11	43.94 ± 13.66	198.89 ± 73.74
	12	103.56 ± 37.92	161.80 ± 56.79

NOTE: Pretreatment values for TCMA = 1.74 ± 0.95 (combined average for adults and children).

It should be noted that with both products the methods employed to assay the residues in either gloves or tee shirts quantified only the parent insecticide and not any breakdown products. Thus it is possible that TCP could have been available for absorption as well as chlorpyrifos and TCMA as well as TCVP; if this were the case, then contributions of these breakdown products could have been made to the urinary metabolite levels observed. It cannot be determined from the current data whether any absorption with subsequent excretion of the breakdown products was occurring, but is certainly a possibility. However, the levels of parent insecticide observed in gloves and tee shirts indicate that the parent compounds were certainly available for absorption.

Summary and Conclusions

Our laboratories have developed protocols for the sampling of transferable residues from the fur of dogs treated with topical flea control products. With the two collars containing organophosphorus insecticides reported upon here, we obtained similar and logical results in the pattern of transferable residues from the fur of the dogs, with the back distant from the collar having the lowest transferable residues and the samples taken over the collar having the highest residues. However, the residues from the TCVP collar were considerably higher than those from the chlorpyrifos collar. These differences in magnitude can probably be attributed to differences in the polymer matrix used for the collar. Because TCVP is of considerably lower acute mammalian toxicity than chlorpyrifos, it is probably not of concern to the pet or to the people that the TCVP migrates out of its collar faster than chlorpyrifos migrates out of its collar. These postulated rates of migration out of the collar are probably the reason that the TCVP collar is only recommended for 4 months while the chlorpyrifos collar is recommended for 11 months, despite the fact that the TCVP collar has a higher percentage of active ingredient than does the chlorpyrifos collar.

We had hoped that the tee shirt data might be a good surrogate for exposure, but there were no useful correlations between the tee shirt residues and the biomonitoring data. The biomonitoring data showed consistently higher residues of urinary metabolites in children compared to adults with both insecticides. In addition to the data shown here, the urinary metabolite data were also calculated corrected for urinary creatinine values, and these calculations did not change appreciably the relationship between adult and children's urine. It appears that little, if any, additional exposure to chlorpyrifos occurs because of the use of this collar, although the data are still somewhat equivocal with respect to children's exposure. However, when comparing our metabolite data to the published NHANES data, the ranges in the geometric means we observed for each age group (children - 8.43 to 19.71 ng/mL urine; adults - 4.79 to 8.58 ng/mL urine) were similar to those in the 95th and 90th percentile of the NHANES data, respectively (*8*). In our studies, we feel that some exposure to TCVP occurred because of the use of the collar. This is further supported by comparing the ranges in the geometric means for our data to the published NHANES data. The range of urinary TCMA in children was 30.84 to 55.25 ng/mL urine, and the range in adults was 12.90 to 20.57 ng/mL urine. Both of the ranges were well above the 95th percentile rankings reported in the NHANES data (*8*). However, because of the low toxicity of TCVP, the significance of this exposure in the risk assessment cannot be concluded at this time.

Acknowledgements

This research was supported by grants from the U.S. Environmental Protection Agency's Science to Achieve Results (STAR) grant program (Grant Numbers R825170 and R828017). Although the research described herein has been funded wholly or in part by the U. S. Environmental Protection Agency STAR program, it has not been subjected to any governmental review and therefore does not reflect the views of the agency. No official endorsements should be inferred.

References

1. American Veterinary Medical Association. *US Pet Ownership and Demographics Sourcebook;* Center for Information Management, American Veterinary Association: Schaumburg, IL. 1997.
2. Ecobichon, D.J. In *Casarett and Doull's Toxicology*, 5th ed.; Klaassen, C.D., Ed.; McGraw-Hill: New York, NY, 1996, pp. 643-698.
3. Boone, J.S., J.W. Tyler, and J.E. Chambers. *Environ. Hlth. Perspect.* **2001**, *109*, 1109-1114.
4. Boone, J.S.; Tyler, J.W.; Davis, M.K.; Chambers, J.E. *Tox. Mechanisms Methods*. **2006**, *16*, 275-280.
5. Chambers, J.E.; Boone, J.S.; Davis, M.K.; Moran, J.E.; Tyler, J.W. *J. Expo. Sci. Environ. Epidemiol.* **2007**, *17*, 656-666.
6. Davis, M.K.; Boone, J.S.; Moran, J.E.; Tyler, J.W.; Chambers, J.E. *J. Expo. Sci. Environ. Epidemiol.* **2008**, URL http://www.nature.com/doifinder/10.1038/sj.jes.7500647.

7. USEPA. *Organophosphate Pesticides in Food – Primer on Reassessment of Residue Limits*; United States Environmental Protection Agency: Washington, D.C., 1999.
8. National Center for Environmental Health. *Third National Report on Human Exposure to Environmental Chemicals;* U.S. Department of Health and Human Services, Centers for Disease Control and Prevention: Atlanta, GA. 2005, NCEH Pub. No. 05-0570.
9. USEPA. *TCVP Registration Eligibility Decision*; United States Environmental Protection Agency: Washington, D.C., 1995.

Indexes

Author Index

Subject Index

D

E

F

G

S

T